Renewable Energy Policy and Politics

Renewable Energy Policy and Politics
A Handbook for Decision-making

Edited by

Karl Mallon

London • Sterling, VA

First published by Earthscan in the UK and USA in 2006
Reprinted 2007

Copyright © 2006

ISBN: 978-1-84407-126-5

Typeset by Composition and Design Services
Printed and bound in the UK by Cromwell Press, Trowbridge
Cover design by Susanne Harris

For a full list of publications please contact:

Earthscan
8–12 Camden High Street
London, NW1 0JH, UK
Tel: +44 (0)20 7387 8558
Fax: +44 (0)20 7387 8998
Email: earthinfo@earthscan.co.uk
Web: **www.earthscan.co.uk**

22883 Quicksilver Drive, Sterling, VA 20166-2012, USA

Earthscan is an imprint of James and James (Science Publishers) Ltd and publishes in association
with the International Institute for Environment and Development

A catalogue record for this book is available from the British Library

Library of Congress Cataloging-in-Publication Data has been applied for

Renewable energy policy and politics : a guide for decision-making / edited by Karl Mallon.
   p. cm.
 Includes bibliographical references and index.
 ISBN-13: 978-1-84407-126-5
 ISBN-10: 1-84407-126-X
 1. Energy policy. 2. Renewable energy sources. 3. Energy development--Environmental aspects.
4. Sustainable development. I. Mallon, Karl.
 HD9502.A2R446 2005
 333.79'4--dc22

          2005027704

Mixed Sources
Product group from well-managed
forests and other controlled sources
www.fsc.org Cert no. TT-TOC-2082
© 1996 Forest Stewardship Council

DEDICATED TO THE MEMORY OF
EMILY CRADDOCK

Contents

List of Figures, Tables and Boxes

Figures

Tables

Boxes

List of Contributors

Karl Mallon trained as an experimental physicist specializing in high energy physics before pursuing an engineering doctorate in renewable energy. He was director Greenpeace Energy Solutions from 1997 to 2001 and worked on energy policy analysis and reform in Europe, North America, Southeast Asia and the Middle East. In 2002 he formed the Transition Institute, an energy think tank and business incubation company based in Sydney, Australia. He is also director of Climate Risk Pty Ltd, a company providing climate change impacts risk analysis for the commercial and residential property sectors. He has provided energy reform expertise to the World Bank, industry bodies and companies in the energy sector, and also several non-governmental organizations. He has also led engineering teams taking renewable energy to the earthquake-stricken region of Gujarat, India, and on the Pacific island of Niue following Cycone Heta. This is first book.

Rakesh Bakshi's achievements in the field of business development in renewable energy have brought repeated international recognition. This includes the Padma Shri, one of India's highest civilian honours; the Prince Hendrik Medal of Honour, for his work in building trade relations between India and Denmark; the British Wind Energy Association's award of the title Wind Energy Pioneer; and the presentation of the 2000 Millennium Award by the World Renewable Energy Network. Mr Bakshi is presently Chairman of RRB Consultants & Engineers Pvt Ltd.

Gordon Edge has been a leading analyst of European renewable energy development for many years and was appointed founding editor of the Financial Times Renewable Energy Report news and analysis service. He became head of Platts renewable energy reporting before taking his current position as Head of Offshore Wind Development at the British Wind Energy Association, with responsibilities extending to wind finance and European relations. Dr Edge entered the business world after an academic career including being the Eastern Electricity Research Fellow in Energy and the Environment at the University of East Anglia in the United Kingdom.

José Luis García Ortega is a graduate in physics (specializing in astrophysics). His research work has included work at the Plataforma Solar de Almería (Almeria Solar Platform) on the Solar Furnace Project. In 1991 he joined Greenpeace Spain as ozone, climate and energy projects campaigner. He established and led Greenpeace Spain's Solar Project and the one of the worlds first Solar School Networks. He is an experienced political analyst and was a delegation member to Conferences of the Parties to the UN Framework Convention on Climate Change in Kyoto (1997) and UNFCCC meeting since then.

Volker U. Hoffmann BSc (Econ.) was born in 1940. He studied economics at Leipzig's university and business school before working as a research scientist at the Institute for Energetics in Leipzig from 1969. His work has been concerned with renewable energy issues since the beginning of the 1980s, and in 1990 he founded the IfE group project on renewable energy. From 1991 he was in charge of the ISE project group in Leipzig, and in 1996 he became project head of the electrical energy systems department and responsible for the evaluation of standards in the German Thousand Roofs photovoltaic programme. He retired from professional life in 2003.

Emilio Menéndez Pérez attained his engineering doctorate in mining, becoming Honorary Professor at Madrid Autonomous University's Ecology Department specializing in the engineering of mines, industrial plants and power stations. Until 2001, he was lead manager of the R&D unit at the Empresa Nacional de Electricidad, SA (ENDESA), Spain's national electricity company. He is a specialist on clean coal technologies and renewable energy development.

Kevin Porter is Vice President with Exeter Associates, a consulting firm based in Columbia, Maryland. Mr Porter has been active in renewable energy analysis and research since 1984 and he is considered one of the pre-eminent national experts on renewable energy. His scope of work and expertise includes the technical and economic status of renewable energy technologies; design and implementation of state and federal renewable energy policies; transmission access and pricing for renewable energy technologies; and electric restructuring in general. He holds a BS in Environmental Studies from Lewis and Clark College in Portland, Oregon, and an MA in Economics from The American University in Washington, DC.

Randall S. Swisher has served as Executive Director of the American Wind Energy Association since 1989. Prior to that, he worked as Legislative Representative for the American Public Power Association and as Energy Program Director for the National Association of Counties. He has also worked as Professional Staff for the House Interior Committee's Energy and Water Subcommittee and as Executive Director for the DC Public Interest Research Group, where he first became involved with renewable energy advocacy in 1975. Between 1976 and 1981, Swisher served as an Adjunct Professor at Georgetown University and Georgetown University Law Center, where he taught courses on energy policy. Swisher has a PhD in American Civilization from George Washington University and a BA in Political Science from the University of Iowa.

Sven Teske holds a Diploma Engineer (Dipl Ing) degree. He was leader of the Solar and Renewable Energy Campaign in Germany between 1994 and 2004. Since 2004 he has been head of Renewable Energy Campaigning for Greenpeace International. Teske is author of a large number of publications in the field of renewable energy and one of the founders of the German Greenpeace Energy Company, a cooperative that sells 'green electricity' building renewable capacity in Germany, Luxembourg and Austria.

Andrew Williamson is a systems engineer and economist working on rural sustainable energy projects in Cambodia and Laos. He is currently the Visiting Research Fellow in sustainable energy at the Cambodian Research Centre for Development. Before moving to Southeast Asia, Williamson was head of the commercial wind development arm of the Sustainable Energy Development Authority of New South Wales, which spearheaded the monitoring, mapping and commercialization of wind resource data in Australia.

Foreword

While completing this manuscript, I have also been providing a client with advice about renewable and low emissions technology development in the Asia Pacific region. The whole world now seems to be worried about how on earth we will reduce global emissions while countries with massive populations, like India and China, cut their path of industrialization. And, on the face of it, the numbers don't look great.

To make a long story short, if we are to avoid dangerous climate change and keep global warming below two degrees, we must limit CO_2 emissions to about two tonnes per person per year across the world by 2050. By contrast, modern industrial economies today have emissions levels from just over 10 tonnes all the way up to 25 tonnes of CO_2 per person per year, and the populations of India and China are coming along quickly.

So the challenge is to have good standards of living, but at a fraction of the emissions. Is it possible?

Well, here is the ray of sunshine. In principle, not a single piece of technology that will be running in 2050 has yet been installed – no computer, no light bulb, no power station, no wind farm, no factory, no farm tractor, no car, no cooker, no fridge, no aluminium smelter. This means that we still have an almost complete choice about what the future of energy will look like and therefore what emissions will be.

Another ray of sunshine is that it is possible to have an industrial economy operating below this emission level with technologies available today and at a very manageable cost. So the window to address climate change is still open and we have a range of commercially available technologies to hand. Personally, I find it exciting to think that fixing such a colossal problem as global warming is within our grasp! I also believe that achieving the level of cooperation needed will represent an important rite of passage for the human race.

But just because we have the technology doesn't mean the problem is solved. On the contrary, that is the starting point; it is the steps that follow that this book seeks to address. As a species, humanity has the opportunity to make a rational choice, but that doesn't mean that we will! Here it's a tug-of-war between the cost of investment in low emissions industries today and the cost of climate change down the track. It may be easy to keep discounting the future on paper to justify inaction, but nature doesn't read. The most coherent argument against investment now for the future is summed up by Groucho Marx, who jested, 'What has posterity ever done for me?'

What this book really does is focus on industries and political systems as they are (not how we might want them to be) and then works out how the to negotiate the policy and political maze needed to get renewable energy into the market. Since only in a handful of countries are renewables reaching their potential, we can truly view it as a world of opportunity, provided we have the right tools. I hope this book serves as a tool box for those readers from government, industry or civil society who are working to unlock the natural energy resources of their countries.

Finally, it is said that wise people divorce their happiness from the outcomes of their toils. To assist with this process in the reader, I offer this comfort which doubles as a caution: whatever happens, the human race appears destined to achieve a low-emission future, either through a new industrial revolution which will include renewable energy, or through economic depletion and collapse, as the costs of climate impacts unrelentingly take their toll. The outcome for the global atmosphere may be the same – the question is whether, as a race, we can get there the smart way.

I extend my sincere thanks to all of the authors who have contributed to this book and I hope that their willingness to share their knowledge provides insight to those who use this book.

Karl Mallon
Sydney
January 2006

How to Read This Book

This book has been structured as a resource book or handbook. It is not a book that needs to be read from cover to cover to be useful; rather, each of the chapters and case studies can be used as stand-alone resources, and each will provide useful insights and tools for the reader.

It should also be stressed that we do not intend this book to be an up to date guide to the status of renewable energy markets or national policies – the market is evolving far too quickly for that, and there are plenty of such resources out there already. On the contrary, herein we have tried to capture what has been going on in important markets during critical phases of their evolution. The knowledge provided by the contribution authors from formative periods is not reflected in the current statistics or legislation; it is information about successes and failure on the way, and the many paths that were taken by multiple actors. Therefore please treat the information here as 'snapshots' from recent history, which may help to interpret the present and inform the future.

Acknowledgements

Much of the knowledge that we are trying to codify in this book has been built up through working with many people over many years. As the editor, I would like to thank all of the people who have worked with the contributing authors in the development of renewable energy in their respective countries. For my own part, I would like to thank the many colleagues who have worked with me in many of the projects and campaigns discussed in this book. In particular I would like to thank Ian Lloyd Beeson, Rick Maddox, Andrew Richards, Libby Anthony, Rob Marlin, Brian Hall and Donna Bolton, Rick Perrin, Philip Clark and Alex Beckett, John Titchen, Megan Wheatley, John Edgoose, Simone O'Sullivan, Steve Beal, Walter Gerrardi, Steven Gilbeaut, Greg Bourne, Melanie Hutton, Brian Hall, Donna Bolton, John Titchen, Michael Vawser, Graham White, Donna Green, Donna Lorenz, David Mair, LJ, Loch, Colin Leibermann, Chris Newbold, Andrew Jones, Andrew Woodroffe, Andrew Post, James Pennay, Simon Molesworth, Robbie Kelman, Ray Nias, Alasdair Lawrence, Julie-Anne Richards, Nic Clyde, Marie Wood, Athena Ronquillo, Ophelia Cowell, Mim Lowe, Edward Broome, Jyotishma Rajan, Kion Etuati, Jan Madsen, Jan Van Putte, Gareth Walton, Erwin Jackson, Jay Rutovitz, Darren Gladman, Alan Graham, Yolande Stengers, Simon Grosser, Rontheo Van Zyl, Peter Lausberg, Peter Cowling, Paul Horseman, Martijn Wilder, David Ryan, Mark Diesndorf, Madeline Cowley, Liz McBurnie, Ken Brown, Megan Jones, Jeremy Shultz, Jenny Mee, Jamie Reardon, Noa Rotem, Arthur Watts, Tristan Edis, Ric Brazzale and Jenniy Gregory, Anna Reynolds, Paul Toni, Alexander Quarles Van Ufford, Caroline MacDonald, Warwick Moss, Tony Trujillo, Iain MacGill, Rob Passey, Muriel Watt, Hugh Outhred, Steven Pritchard, Stewart Eagles, Gillian, Peter and Anne Tedder, Steve Sawyer, Corin Millais, Bill Hare, Kirsty Hamilton, Ian Reddish, Martina Krueger, Lyn Goldsworthy, Janet Dalziell, Ben Pearson, Anthony Froggatt, Catherine Fitzpatrick, Steven Gilbeaut, Tarjie Haaland, Bernard Huberlant, John Walter, Danny Kennedy, Jan Madsen and Nick Clyde, Jan Van De Putte, William Hobson and Jon Waller, Arthuros Zervos, Soren Krohn, and Gareth Kershaw.

I would like to thank Janice Wormworth for providing fresh eyes and a diligent edit of all chapters.

I would also like to acknowledge the endless support of my upstream family Brenda, Styx, Crystal and Danny, and downstream family Zoot, Django and Quinn. And most of all my fellow stream swimmer Ruth Tedder.

List of Acronyms and Abbreviations

ACNT	Australian Council of National Trusts
ADB	Asian Development Bank
ARG	Australian Research Group
ASEAN	Association of Southeast Asian Nations
AusWEA	Australian Wind Energy Association
AWEA	American Wind Energy Association
BIS	Bureau of Indian Standards
CASE	Commission for Additional Sources of Energy (India)
CDM	Clean Development Mechanism
CPUC	California Public Utilities Commission
CRCD	Cambodian Research Centre for Development
C-WET	Centre for Wind Energy Technology (India)
DNC	declared net capacity
DNES	Department of Non-conventional Energy Sources (India)
DNO	distribution network operator
DTI	Department of Trade and Industry (UK)
EAC	Electricity Authority of Cambodia (Cambodia)
EDC	Electricité du Cambodge (Cambodia)
EFL	Electricity Feed Law
EIA	environmental impact assessment
EPIA	European Photovoltaic Industries Association
ERCOT	Electric Reliability Council of Texas
EU	European Union
EWEA	European Wind Energy Association
FDI	Foreign Direct Investment
FFL	Fossil Fuel Levy
FIPB	Foreign Investment Promotion Board (India)
GDP	gross domestic product
GHG	greenhouse gas
GM	genetically modified
GMO	genetically modified organism
IDAE	Instituto de Diversificación y Ahorro Energético (Spain; Energy Diversification and Conservation Institute)
IEA	International Energy Agency
IIP	International Industrial Partnerships
IP	intellectual property
IPPs	independent power producers
IREDA	Indian Renewable Energy Development Agency Ltd
ISO4	Interim Standard Offer #4
ISCC	Integrated Solar Combined Cycle
IT	information technology

JICA	Japanese International Cooperation Agency
LEDs	light-emitting diodes
LETAG	Lower Emissions Technical Advisory Group
MIME	Ministry of Industry, Mines and Energy (Cambodia)
MNES	Ministry of Non-conventional Energy Sources (India)
MOE	Ministry of Environment (Cambodia)
MRET	Mandatory Renewable Energy Target
NEDO	New Energy and Industrial Technology Development Organization
NETA	New Electricity Trading Arrangements
NFFO	Non-Fossil Fuel Obligation
NFPA	Non-Fossil Purchasing Agency
NGO	non-governmental organization
NI-NFFO	Northern Ireland Non-Fossil Fuel Obligation
NIS	National Institute of Statistics (Cambodia)
NPBD	National Project on Biogas Development (India)
NPIC	National Programme on Improved Chulhas (India)
NREL	National Renewable Energy Laboratory
NRSE	New and Renewable Sources of Energy
PTC	Production Tax Credit
PUC	Public Utilities Commission
PURPA	Public Utility Regulatory Policies Act
PV	photovoltaic
R&D	research and development
RCEP	Royal Commission on Environmental Pollution
RD&D	research, development and deployment
RE	renewable energy
REAP	Renewable Electricity Action Plan
RECs	Renewable Energy Credits
REEs	rural electricity enterprises
REF	Rural Electrification Fund
REL	Renewable Energy Law
RETs	Renewable Energy Technologies
RO	Renewables Obligation
ROCs	Renewable Obligation Certificates
RPS	Renewable Portfolio Standard
SME	Small and Medium Enterprise Cambodia
SO	System Operator
SRO	Scottish Renewables Obligation
UCS	Union of Concerned Scientists
UNFCCC	United Nations Framework Convention on Climate Change
WB	World Bank
WCD	World Commission on Dams
WWF	World Wide Fund for Nature

1

Introduction

Karl Mallon

---

*Where there is a conflict between an available clean technology
and an entrenched dirty one, the challenge is politics
and the need for legislative action, not technology...
We can do it, we just have to want to.*
DAVID SUZUKI (Canadian scientist, environmentalist and broadcaster)

With the Montreal climate change meeting it finally appears that the scientific
argument for action to address anthropogenic climate change is essentially complete. The debate is no longer if climate change is happening, but how quickly.
Scientists are mounting increasingly urgent warnings about climate change, policy-makers are casting about for ways to slash emissions, and the need for effective
renewable energy policy-making is becoming ever more important.

This book starts with the assumption that renewable energy is indispensable
to the 60–80 per cent greenhouse gas (GHG) emissions cuts scientists say must
be made over the next 50 years. According to the IEA (1999), 'The world is in
the early stages of an inevitable transition to a sustainable energy system that will
be largely dependent on renewable resources.' This may not be the starting point
for some readers who may also be considering sources such as nuclear power and
the geosequestration of coal-fired power station emissions. We will not have that
debate here, but I do invite such readers to ensure they are familiar with the regional
stability issues of plutonium stockpiles if the urgent needs of Asia for energy are
met through nuclear power, and also the impacts of geosequestration on the cost of
coal-fired generation versus projections for wind and biomass. However, what we
can perhaps agree on is that the time for continuing to build more plants that push
carbon into the air is largely over and renewable energies are the best thing we have
so far – they are safe, secure, scalable and sovereign forms of energy supply. If they
meet all our future needs in due course that is great, but if they do nothing more
than buy time while better technologies are evolved that is also good. Either way,
the urgent need for renewable energy development remains the same.

Given the manifold benefits, many might ask why renewable energy technologies cannot find their own place in our energy systems spontaneously and

without the assistance of policy. Unlike mobile telephones or early desktop computers which created new services, renewables are entering an existing market which is already occupied by another product, namely fossil fuels. The incumbent is cheaper and in some ways more flexible, and price is – for the moment – the universal driver in free markets.

Nonetheless, the cost of renewable energy continues its downward trajectory. The Kyoto Protocol's February 2005 entry into force will only aid this decline, as previously externalized costs of greenhouse pollution are increasingly factored in. The price of renewable and conventional generation will continue to converge; like a pack of jackals hunting down the fossil fuel wildebeest, the conclusion of this confrontation appears inevitable. Ultimately it will be hard for fossil fuels to compete with technologies that make energy from fresh air or sunlight.

Until this occurs, however, expecting renewables to compete on an equal price footing with well-established conventional technologies is a tall order. Instead, policy is needed now to unlock renewable energy markets and establish the expertise and manufacturing that will deliver the medium- and long-term deep emissions cuts that are urgently needed.

Given the need for policy changes, one might expect their implementation to be a fairly straightforward exercise. Indeed, looked at in the cool light of logic, energy and environment issues are technical debates with clearly right and wrong answers that one would expect to have an objective of social, health, economic and environmental benefits. Theoretically too, policy and legislation are rational processes which aim to improve the wellbeing of society and reduce the exposure and risk of external threats to that wellbeing.

In reality, however, political decision-making is often far from rational, and energy policy is no exception. Conventional energy businesses are corporate colossi that may not countenance new industries that encroach upon or even question their dominance. I have known mining giants and oil majors seek to have even the smallest renewable energy schemes strangled at birth. Unwitting renewable energy industry participants and their allies are lambs to the slaughter unless they realize the magnitude of the challenge that faces them.

This book seeks to prepare renewable energy proponents for the many skirmishes they will encounter on the path to sound renewable energy policy. We will learn how to shatter the old myths that tell us that technical breakthroughs, and not market development, are the solution. We will discover how to cut across the voices chanting, 'Let free markets do the work'. We will equip ourselves with the economic answers for tight-pursed finance ministers who demand, 'Why ask for price support if renewable will become cost competitive some day anyway?'

To understand our goals and learn from those who have tried before, we will perform a full 'physical exam' on policy measures that failed in the past, to determine what went wrong. For example, what missteps can lead well-intentioned renewable energy policies to subsidize new, gas-fired power plants or even coal plants instead of renewables? We will come to understand the characteristics of sound renewable energy policy, such as schemes that create a healthy, steady pull on industry. We will also learn how to avoid policy that prompts a development gold rush only to leave behind an industry ghost town.

Once we understand good policy, we will move on to politics: the crucial task of negotiating renewable energy policy change. We will learn to identify who is on the renewables team, who the opposition is, and who is sitting on the bench (or the fence). We will also walk through important steps in actual campaigns for policy change, for example, how to maximize the positive space for political decision-makers to make positive changes.

By way of demonstration, this book also walks the reader through real-world case studies of energy policy success and failure in the United Kingdom (UK), United States (US), India, Spain, Germany and Cambodia.

For the UK, case study author Gordon Edge highlights the dangers of creating a renewable energy driver without preparing an integrated policy framework to deliver it. Noble aspirations of new technology development were pursued in a bidding environment so competitive that new projects were marginalized to the extent that they were not even built.

Randall Swisher and Kevin Porter will walk us through US policies that achieved early meteoric successes, making California a world leader in wind installations, only to be withdrawn as oil prices unexpectedly declined. The authors also untangle the policy triggers of repeated boom–bust cycles that continue to this day. And they show us the secrets of success for the Texas Renewable Portfolio Standard, an aggressive promoter of wind power.

For India, case study author Rakesh Bakshi presents us with the challenges of a rapidly developing nation with an inadequate and erratic power supply. With what could become the largest power market in the world, the gap between consumption and supply is widening, and polluting emissions are growing. In the face of these challenges we will see how India has become one of first countries in the world to assign a ministry to the development of renewable energy.

In the case study on Spain, José Luis García Ortega and Emilio Menéndez Pérez describe how pressure from civil society groups for renewable energy actually materialized before, and then drove the development of, a domestic industry. We will see how bold aims for environmental, employment and energy security have largely been achieved, catapulting Spain to a leadership position in the global wind industry, and making it a major exporter of solar photovoltaics (PV).

On Germany renewables policy, Sven Teske and Volker Hoffmann will describe how a nation now at the forefront of solar PV development experienced a rocky road to success. They will take us through a roller coaster of federal policy changes that kick-started a vital industry only to leave it on a meagre life support system of lower-level measures. We will see how this disarray essentially amounted to an unplanned five-year, de facto trial of renewable energy policies across the various German states.

And finally in Cambodia, Andrew Williamson will show us the forces that drive renewable energy policy in one of the world's 20 least developed nations. We will see how policy in such countries can have more to do with the politics and goals of international development agencies than the priorities of the resource-strapped governments themselves. We will also see how forces such as technology stigma create subtle but effective barriers that conspire against the use of clean energy technologies in Cambodia.

It is my sincere hope that this book will provide you, the reader, with some tools and insights to further your own particular quest for renewable energy development, whether you are from the ranks of industry, government or civil society. If you are successful, we may have a world where energy passes through our lives like water, used and enjoyed but without a deadly environmental penalty attached. But please act swiftly and we may avoid the worst excesses of greenhouse gas pollution and its consequences being locked into the atmosphere.

Reference

International Energy Agency (1999) *The Evolving Renewable Energy Market*, IEA, Paris

2
Myths, Pitfalls and Oversights

Karl Mallon

Introduction

In this chapter we take a walk through a world of renewable energy policies to differentiate successes from failures. In fact, the intention here is to get as much policy failure onto the table as possible. Although I considered starting the book from a more positive vantage point, I decided instead that taking aim at failures would clearly identify the issues we need to tackle. Thus, this chapter identifies what must be avoided if we are to walk the path to successful renewable policies. The country chapters that form the second half of this book provide many illustrations of the issues we will cover and so I will contain the discussion points to the question rather than repeating examples.

I have grouped the discussion into three sections. First, we cover some of the myths often found at the root of poor policy. Second, we examine the pitfalls by which a well-intentioned policy fails to achieve its objective. Finally, we scrutinize oversights wherein policies that should have been established were overlooked.

Myths

The technical myth: The price of renewable energy will be reduced by technical breakthrough

The first myth of energy is that some great idea will come along and revolutionize its production, making energy cheap and clean, allowing man to live in peace, enriching the poor and healing the lame. We have heard this about nuclear power, fusion reactors, solar power and many other technologies. For example, several times a year a claim rings out about a new technology that will make solar power affordable to everyone. But as we hear from KPMG, the only thing required to make solar affordable is market size:

> *Even in a relatively small, cloudy and rainy country such as the Netherlands, there is an enormous market potential for solar panels. If this entire potential were utilized, solar energy could provide three-quarters of the Netherlands' electricity needs.*
>
> *The size of the current market for solar panels stands in stark contrast to the potential of solar energy. Up to the present only a fraction of the possibilities have been utilized. The most significant reason for this is the fact that the price of solar energy is much higher than the price of conventional energy. In the Netherlands the price difference is up to a factor of four to five. The predominant reason for this is that the demand for solar energy and solar panels is small and the associated prices are high. It comes down to a classic chicken/egg problem: as long as demand is small, production of solar energy will remain smallscale and expensive, and as long as the production is small-scale and expensive, the price will remain high and the demand small: catch 22.* (KPMG, 1999)

Of course continued innovation will ensue, but it is market growth that will drive and deliver that innovation, not the other way round.

In fact, most renewable energy sources have had their big breakthroughs, and are now on the path of development trod by every other technological breakthrough. Once the initial technology is proven, the commodity price is driven by the size of the market. The typical behaviour is one in which a doubling in installed capacity has an accompanying 20 per cent decline in the commodity's price.

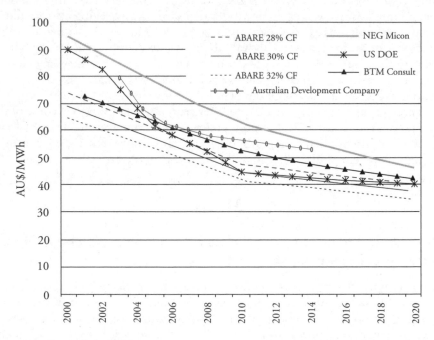

Source: Transition Institute (2004)
Note: CF = capacity factor

Figure 2.1 *The forecast cost decline of wind-generated energy in Australia*

Waiting for further technical breakthroughs puts the cart before the horse, and in so doing distracts policy-makers from addressing market development issues that actually underpin the technical evolution and commodity price.

However, these myths have an excellent pedigree; they do not emerge from out of the blue. The industrial revolution was driven by a technical breakthrough, which was the steam engine used to harness an energy-intensive fuel source – coal. In a similar way, the internal combustion engine was a new technology applied to a 'new' highly concentrated fuel source – oil. Nuclear power was another technical breakthrough which unlocked the new fuel source of uranium.

Thus, since the industrial revolution, our view of progress has been tied to the idea of a breakthrough. However, with all but a few exceptions, most progress comes from combinations of steps which, when taken as a whole, appear as a breakthrough. A modern Mercedes Benz that does 300kph is not the result of the internal combustion engine breakthrough. It is the result of a century of steam engine design progress, then the combustion engine breakthrough, followed by a century of incremental development which has been paid for by ever-expanding car sales.

Consider that the photovoltaic (PV) effect was discovered by Edmond Becquerel in 1839. It took 120 years for Bell laboratory scientists to come up with the material that would unlock sunlight as a small source of electricity. It was possibly the first example of generating electricity without requiring physical movement – a truly remarkable invention! Yet so ferociously expensive was it that only the space industry, pushed by their need for power on satellites, could afford to use it.

Yet one niche opens up another. PV for satellites led to land-based remote power supplies. When demand goes up, the price comes down. The grand industry for land-based power production using solar panels can be credited to people such as David Katz in the US, who began hooking up solar panels to Russian submarine batteries to deliver basic energy services to off-beat, off-grid communities in the Californian hills. From these quirky beginnings we now have major industries in Japan and Germany manufacturing thousands of rooftop PV power stations.

Notwithstanding some surprises that may yet come, we already know the main primary renewable energy sources. We can also measure the amount of raw energy available to be harnessed from these sources. For example, we know the physics of the wind and how much energy is contained in a given catchment area. Similarly we know the energy available in each photon (particle) of light and how much sunlight the sun sheds on the Earth each day. We know how much energy is in a wave, a tide, a river's flow or a kilogram of organic matter.

We have developed technology that can harness the energy of many of these sources with varying efficiencies. Some of these technologies are on cost trajectories to commercial viability compared to 'conventional' energy, so these are the technologies that we might consider to have 'broken through'. This list includes wind, solar hot water, solar PV, biomass and derivatives, small hydro and geothermal. For others we currently lack the technology that will definitely do the job at a commercial scale and be capable of deployment in large numbers, so these sources can be said to be still working on a breakthrough. Here we might include

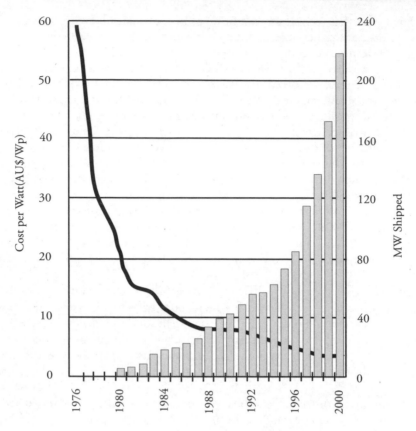

Source: EPIA (2003)

Figure 2.2 *Solar photovoltaic market growth*

wave, tidal, ocean thermal and ocean current sources and a wide array of solar thermal electric technologies (no pun intended). Yet even for those sources that cannot yet be harnessed commercially, we at least know what volumes of energy to expect if successfully harnessed. So we can anticipate fewer surprises from renewable sources than we might be led to believe.

Some renewable technologies have even evolved beyond the dominating drives for efficiency and low cost. For instance wind turbine manufacturers elect to use slightly less efficient three-bladed machines instead of two-bladed machines because they have a better visual aesthetic (some say because they look more like flowers!) and also tend to optimize for minimum noise rather than maximum power.

To summarize, the recognition of both commercial and pre-commercial renewables clearly implies the need for two very different policy types. And the myth that all renewables are waiting for a breakthrough to become cheap is damaging and counter-productive. For commercial renewables, persisting with research and development (R&D) funding instead of market development is akin to trying to put nappies on a teenager.

The myth of the righteous: A good idea will always succeed and intervention is unnecessary

Most significant innovations that have led to major societal changes have been identified and fast-tracked by governments. This is especially so if the benefits are not directly commercially valuable.

Unfortunately the road to oblivion is littered with good ideas. Perhaps, as a society, we assume that because we are able to determine what is best, we assume that best will triumph. This belief may hold true in the laboratory, but it quickly breaks down in the world of powerful business and politics. For ideas and innovations to succeed the support of the business and political realms is critical – they must feel these innovations are good for them. In general a 'good' thing must either do something that could not be done before, or do what can already be done a lot better. And to judge that we usually use price.

When an innovation opens up new horizons, its perception as good is almost universal. The desktop computer is a good thing, because there was no equivalent before it. Cheap air travel is a good thing because it opened up a world of new possibilities for people, industry and business. The World Wide Web is a good thing, again because it opens up a whole new space.

However, most profound innovations in our modern life have largely been identified and had their development assisted in some way by governments, often the military. In the last century, modern civil aviation followed massive military spending during the Second World War, which developed low-cost, reliable, high-capacity aeroplanes. The computer industry is again a product of the Second World War, and massive US government spending thereafter; now US-born companies dominate the global information technology (IT) economy. The nuclear industry is another example. Indeed, its civil power application faltered in both the US and the European Union (EU) in liberalized markets.

Today we see governments fast-tracking genetics, nanotechnology and indeed renewable energy. We do not see those governments who intend to capture a new market standing by and saying, 'If it's a good idea let it sink or swim.' There is a recognition that industrial economies are built on innovation, and so we see certain countries grasping renewable energy technologies and seeking to establish their place in a large future market. It comes as no surprise that it is the 'innovation' economies that are leading the way.

Although there may be almost universal agreement that renewable energy is a good idea, this agreement is not sufficient to foster a substantial industry. For specific renewable energy technologies to succeed, pointed and well-conceived government intervention is required. This intervention mainly takes the form of legislation.

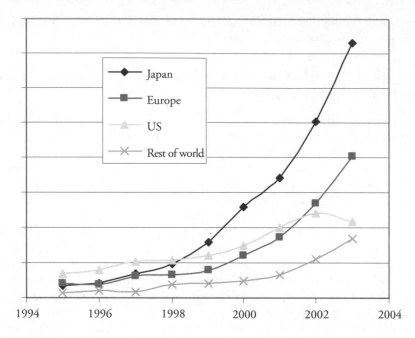

Source: EPIA (2005)

Figure 2.3 *The race to lead the global solar electric market; the countries who have built their economies on innovation are leading the way*

The hands-off myth: Government intervention only undermines the proper working of markets

Markets are good at doing what they are designed to do. They cannot do what they are not designed to do. It is often said that they are good servants, but poor masters. Governments have a fundamental role, perhaps responsibility, to establish the market conditions.

Is email a good innovation? Not if you sell fax machines. Was the fax machine a good idea? Not if you sold stamps. In fact the fax machine was better than letters in various regards and email is better than a fax machine in all but a few ways. However, 'better' can be highly subjective.

While renewable energy sits in the realm of the better, its position there is not unequivocal. Renewable energy is better than conventional generation because the fuel is free, inexhaustible, less polluting, is generated indigenously, and because it creates more employment, among other reasons. However, it can be considered inferior to current conventional supplies and systems in that some renewable forms are less energy intensive, the sources are nature-dependent and not as controllable, and some are harvested on a smaller scale which may increase complexity.

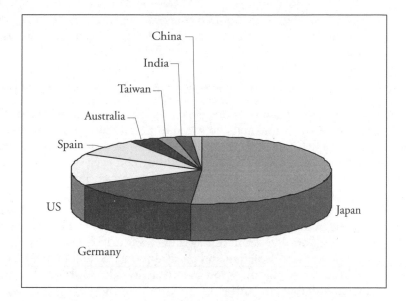

Note: The pie chart for sales to the end of 2003 shows Japan and Germany have secured dominant positions in the global PV market, with the US early lead being lost due to lack of government market development intervention.
Source: EPIA (2005)

Figure 2.4 *Developing countries are exploiting an opportunity to take a leading manufacturing position in the solar technology from the outset, with India and China leapfrogging to the fore of the global PV industry*

With renewable energy the notion of 'better' depends on the relative importance placed on the various advantages and disadvantages. For renewables many of the advantages sit within the social and environmental arena while the disadvantages sit in the purely business and industry sectors.

In a standard power market, must the powers that choose technologies consider the social or environmental impacts of pollution? No, the taxpayer and insurance companies pick up that tab. What about the effects of carbon emissions? No, this pollution is not geographically specific and its costs are deferred to future generations and governments. As for the benefits to the environment, employment and industrial development, these accrue at a national or regional level outside the power sector. Since it is not possible for Swiss Re or the government of Tuvalu to set energy pricing and environmental loading around the world, or even sue for damages, another bridge must be found.

Why would a sector adopt a technology if the benefits offered did not accrue to the businesses involved? It would be financial dereliction to do so. Therefore, for business to act, there must be benefits or penalties set up to create an incentive for change.

Governments aiming to harness the benefits of renewables must intervene because the power sector and energy markets lack ways to incorporate the benefits into their decision-making. The required intervention must create price signals that allow companies to reflect the external benefits in their balance sheets and thereafter leave markets to maximize efficient delivery.

Money myth: Renewable energy is more expensive than thermal power

Current renewable energy is only more expensive than thermal (nuclear and fossil) generation if the environmental and social impacts, the 'externalities', are not priced. Failure to acknowledge this in some way leads to distorting policy frameworks. Furthermore, renewable prices are declining and even in the most hostile markets they will continue to converge in price with conventional energy sources without externality pricing.

The myth of cost efficiency is really a realm of 'double-think', the Orwellian place where a person maintains two contradictory beliefs. The acknowledgement that air-borne pollution does damage has long been driven home, as have the impacts of acid rain, ozone depletion and persistent organic pollutants. However, it is still possible for society, industry and governments to know this and yet ignore the social and environmental damage in the cost assessment. Not only is it still possible, the practice is almost universal in the energy sector.

Until polluters are required to avoid or compensate for the secondary effects of their business, business planners will not consider these costs on their balance sheet. However, once legislative controls are in place, the financial impacts of the former externalities are acknowledged, and business planners are forced to factor

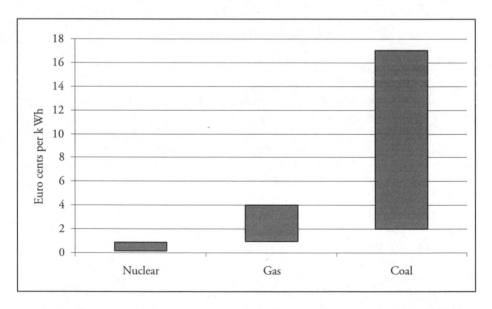

Source: European Extern E project, as presented in EWEA (2004)

Figure 2.5 *The external costs of thermal power generation in the EU*

them in. Thus the triple bottom line condenses into the single bottom line, which is the only aspect most investors pay attention to anyway.

Until this 'internalization' of externalities is accomplished, it is standard practice to ignore pollution. Society and the economy pick up the tab and are unwitting subsidizers of industry. The big problem for renewables, however, is that externalizing these costs causes market distortions that prevent more fully cost-effective technologies from entering the market.

Elsewhere in this book, we too will play the double-think game and refer to costs of different energy types *without externality costs included*, barring the occasional examination of renewables' impacts on the emerging greenhouse gas (GHG) emissions trading markets. This is to avoid confusion and also to avoid having to refer to or decide on the different estimates for externalities. However, for the record, we will lay out the best estimates of these externalized costs in Figure 2.5 and suggest the reader makes a point of adding them into any comparisons they may find themselves making.

Box 2.1 *Internalizing the external costs of radioactive pigeons*

The most bizarre illustration of externalized costs I ever encountered was a 1998 incident concerning pigeons and the Sellafield nuclear reprocessing facility in the UK, a plant infamous for its radioactive emissions. It appears radiotoxic emissions were gradually 'charging up' the pigeons that frequented the site to the point that they could be classified as low-level radioactive waste. What is more, these pigeons were flying around the local town dropping their own waste, also classified as low-level radioactive. And so a crack squad of radioactive pigeon catchers was set into motion and a radioactive bird poo van dispatched for washing down window sills and car bonnets. In this case, if the council paid for the clean-up van, the costs were externalized and, if the company paid, they were internalized.

Money myth: Because renewable energy is more expensive and capital intensive, the best governments can do is to throw money at the problem

As each renewable energy technology becomes more commercially mature, governments become less significant as providers of the direct capital support needed to make up the cost difference relative to conventional generation. What governments can do instead is focus on policy design and legislation to attract private sector investment.

Today a typical wind turbine costs about €2 million to buy and install. A wind farm in the more developed markets has a capacity of 50 megawatts (MW) and more in total capacity (one turbine generates approximately 2MW). Calculations I have been involved in recently indicate that a manufacturer needs to sell about 100 turbines per year to justify a blade manufacturing facility. A market with the capacity to support just one manufacturer is not an efficient industry.

Strong markets are usually characterized by three or more competitors, with no individual company having more than 30 per cent market share. This suggests a strong installation and manufacturing industry might be characterized by 300 turbine sales per year, equating to sales of €600 million per year for a healthy national wind industry.

It is clearly beyond the budgets of most governments to directly inject money into renewables in order to fast-track a competitive industry. A handful of demonstration projects might be useful, but this does not spawn an industry.

Furthermore, why put up government money when it can come from other sources? Individuals in investment circles repeatedly point out that the world is awash with capital, but there is a shortage of projects. The shortage is even more acute for environmentally sound projects sought by ethical investment houses. This shortage of projects warns us that conditions in many countries are not right for investment. This is where governments can act most effectively.

Governments' critical role is their use of legislation and market dynamics to leverage private sector investment into renewable power projects and industries. Past successful policies provided secure markets for projects by legislating to distribute the premium costs of renewable energy development onto the energy consumer (who is ultimately responsible for the source of the pollution in the first place) or onto the taxpayer (whose natural and social assets are being protected). Wide distribution of this cost ensures that its price impact is minimized. We will see practical examples of success and failure in the country chapters later in the book.

Policy Pitfalls

Pitfall: Under-defined objectives

In mid-2003 I was asked by the World Bank (WB) to attend its review of the Bank's activities in extractive industries. Extractive in this context refers to extracting resources from the ground – from mining gold to drilling for oil. The Bank's lending to extractive industries and projects has been under pressure from many sides, from indigenous people unhappy with intrusive developments, to international pressure groups calling for a halt to fossil fuel lending. I was asked to speak about some of the alternative options in the energy field (World Bank, 2004).

In one particular workshop a heated debate sprang up about misspent funds, and the issue of companies and politicians lining their own pockets. People were upset about the failure of the money to get through to ordinary citizens or indeed the national economy that services the population. If corruption existed in a country, was it the Bank's responsibility or the citizens'? How could good or bad companies be identified? How could the Bank ensure that project profits were not concealed by the company? And so on.

All of this was most engaging and it was difficult to allocate responsibility or blame, until the objectives of the Bank were examined. The Bank's aims in lending as set out in the meeting came back to twin objectives of promoting sustainable development and eradicating poverty. But if those were the objectives, why were they not spelled out in the contracts? The structure seemed to dictate that the World Bank must fund projects rather than establish contracts for investment to achieve its actual desired objectives.

The idea of funding one task while assuming another will follow is a dangerous guessing game to be avoided at all costs. If you want to eradicate poverty then take out a contract with someone prepared to do that, and if their means of doing that cost-effectively is by mining gold, then well and good. But it must be acknowledged that the mine is only an intermediate step. Lending money for a gold mine and simply hoping that this will eradicate poverty is poor contracting. The end objectives must be clearly understood by all parties throughout the process and where possible hard-wired into contracts/legislation.

This message is equally pertinent to the renewable energy industry. We must be very clear that the objectives we have for renewable energy and industry are explicitly embedded in the policies legislated. If we want definite outcomes, then the more specific our policies regarding these outcomes, the better. The following quotes from our case study authors illustrate this. In the first, Edge describes a shortcoming of a policy called the Non-Fossil Fuel Obligation (NFFO) in the UK during the 1990s:

> *In common with other European governments the UK was coming out of a period of very high unemployment when NFFO was being set up. Crucial vote-winning issues were new industries and new jobs. The establishment of a UK technical and manufacturing lead in these new industries and technologies was much discussed, but failed to be expressed explicitly in the policy make-up of NFFO. This may have been a function of the then government's faith in free market economics.*

A provision for manufacturing or local content may or may not have been appropriate for inclusion in the NFFO mechanism itself, though EU state aid rules would likely have precluded an explicit provision. However, since it appeared nowhere in the supporting policy frameworks, the ability to ensure the delivery of manufacturing and employment was left beyond the control of government – a significant omission given that the government held the purse strings.

In a second example, García and Menéndez describe the role of Spanish states in successfully assuring their stated outcomes:

> *Furthermore, the states have undertaken aggressive targets to push activity along. For example, Galicia, which has an Atlantic coastline, has had a 2300MW wind target since 1997 – equivalent to 45 per cent of the state's power demand. This has been tied to the aim of ensuring that 70 per cent of the investment is spent inside state borders. It has resulted in over 5000 direct and indirect jobs and numerous factories.*

Pitfall: Cross-cutting objectives

Sometimes apparently mutually supportive objectives can clash in practice. The political proponents of policy change must balance economic, industrial, social and environmental interests to define a clear set of policy systems that will actually deliver intended outcomes.

Some common objectives of renewable energy promotion include industry development for employment, securing an export position in emerging markets, security of energy supply (reducing fuel or electricity imports), long-term GHG emissions reductions and increasing rural economic development.

Some common areas of cross-cutting objectives occur when a one-size-suits-all approach is taken. The elegance of simplicity is often attractive, but in renewable energy policy it can lead to an inferior outcome.

A key example occurs when renewable energy development is cross-cut with carbon abatement. In principle, carbon trading entails applying a cap or pollution permit requirement to carbon, increasing the cost to carbon polluters and thereby closing the gap between conventional and renewable generation prices. However, if carbon trading allows for the use of tree planting (carbon sinks) or displaced emissions to be included, very low-cost carbon abatement is possible.

Nevertheless, low-cost abatement measures are limited and even those that are available are not necessarily secure. In the longer term it is generally agreed that renewables are most likely to provide the least-cost abatement for deep cuts in GHG emissions. The policy pitfall in the short term is that we may hobble the current renewable industry by throwing it into a scheme where it is required to compete with low-cost carbon abatement. And in so doing, the opportunity for low-cost abatement in the future is compromised.

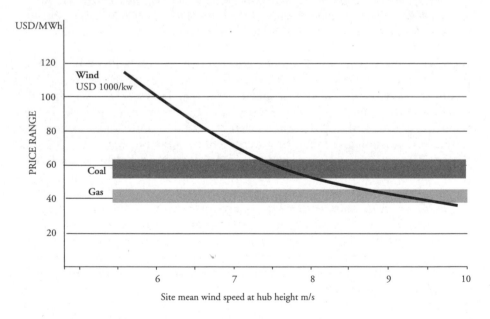

Source: Windpower Monthly (2004)

Figure 2.6 *Wind energy can be competitive with conventional sources, as shown here for the EU, but as carbon is priced the coal and gas bands are pushed upwards*

Box 2.2 *Policy failure in the fictional state of Djangostan*

Imagine a fictional scenario in which a policy-maker is asked to ensure his country, let's call it Djangostan, will have a clean energy industry with plenty of jobs, reductions in GHG emissions, and cheap, domestically produced energy. This Djangostani official drafts a policy that provides a price for power that renewable energy developers would find attractive. Then a delegation from Canada suggests the policy should not be so technology-specific, why not define it in terms of zero-emission technology? So the Djangostan policy-maker includes nuclear power in the scheme. Next, a gas pipeline builder drops in and argues that gas is the cheapest form of emissions abatement in the generation sector. Then a coal lobbyist comes by and talks about the latest coal technology, its efficiency, and the future of carbon capture and sequestration. And wouldn't it be best to let the market decide? So the policy-maker drafts a technology-neutral carbon trading type policy.

What Djangostan gets as a result is a group of Djangostani farmers who are promising not to clear forested land to anyone who will pay one dollar per tonne of carbon dioxide (CO_2). A new coal power station manufactured in the US that was going to be built anyway, but because it is more efficient than the old one, carbon credits are created at almost zero cost. Developers for the gas pipeline and generation facility, which would have been a little better, do not bother coming because the now-subsidized coal makes it uneconomic. And finally the renewable energy companies that were looking to set up in Djangostan pack their bags and go elsewhere. Building a renewable energy plant is a real investment and they cannot compete against the smoke and mirrors of 'avoided' emissions in coal plants and trees that 'might' have been cut down.

No renewable energy is installed. No factories are set up. No extra people are employed. No fuel imports are displaced. All that has happened is what would have happened anyway, except that some money changed hands and some politicians and business people said they were protecting the planet. And of course some farmers are smiling broadly.

What went wrong? First, the objectives were not prioritized. The objectives of carbon emissions reductions and least cost came to dominate policy construction, taking precedence over industry development and indigenous supply.

Second, the time frames were not specified. No industry starts out cheap, so decisions need to be made about the time frames for each of the policy objectives. Long-term economic efficiency and economic development may well have led to different policy choices. Is it more cost effective to be a manufacturer and exporter or an importer of a given technology? What are the co-benefits worth striving for? What cost to the economy spread over what period is acceptable? These are some of the questions that needed to be asked and answered at the outset, not in retrospect.

This is not necessarily a policy failure as regards GHG emissions, but it is with respect to renewables. It shows the need to adequately identify, prioritize and separate policy outcomes.

Even for GHG emissions reductions the catch-22 discussed in the introduction emerges again. In order to achieve low-cost emission abatement, one must first put in place industry development.

Pitfall: Inadequate resource and/or technology identification

'Technology neutral' is a buzzword often heard in climate and energy circles. It is meant to imply a white-gloves, optimum-outcome approach that does not prejudice or favour a particular technology. However, a wide variety of technologies can all claim to be green, reduce greenhouse gases or be zero emission. In fact, there are real risks inherent in policies that fail to clearly define which technologies, groups of technologies or resources they support.

There are three main reasons technology neutrality should be avoided: to omit out-of-scale projects and free riders; to keep out dead-end innovation; and to avoid killing off future industries through excessive competition.

Out-of-scale projects and free riders

The need for focused identification is easily illustrated when we examine the relative scales of industries and technologies we are trying to leverage. In a new national market for clean energy, even large-scale renewables like wind and biomass, which are likely to produce projects 10–50MW in size, might be expected to yield only a handful of projects annually in the first few years. By contrast a large hydro project or high-efficiency coal project would have a capacity of 500–2000MW or more. So this single conventional project could soak up the 'support' for the equivalent of a dozen of renewable projects.

Furthermore, there are serious questions about additionality when other technologies are included. By additionality we mean that the project would not have happened without the new policy, especially if it is conventionally fuelled. For example, a new coal plant may be 5 per cent more efficient than the national average, but would that machine not have been installed anyway under business as usual? If so, under an additionality clause, it would be excluded from claiming emissions reductions credits.

Broad definition has killed the focus of renewable energy policy packages in some US states, as an excerpt from Swisher and Porter's US case study points out:

> [What made the Texas Renewable Portfolio Standard (RPS) work was a] careful definition of the renewable technologies. By contrast, other states such as Maine had such a broad definition, including a number of non-renewable technologies, that the state RPS had virtually no impact on the market for renewables.

Dead-end innovation

The hydrogen-from-coal story (see Box 2.3) makes the point that we can sometimes entertain change that does not in fact bring any positive progress. This example examines the idea of fossil fuels being used to make hydrogen. Given that this technology would provide no net decrease in GHG emissions, what would be the point? This is dead-end innovation.

Box 2.3 *Hydrogen from coal as dead-end innovation*

Hydrogen is getting a pretty good name as the fuel of the future, and indeed it should. I remember one promotional automobile advertisement that depicted a glass beneath a tail pipe, half full of water. True enough, the oxidation of hydrogen releases ample energy and the chemical result is H_2O.

Yet we must ask, 'Where did the hydrogen come from?' In Germany, where a famous car manufacturer trialled its fleet of hydrogen vehicles – and promoted the scheme with photos of a glass of water filled from the exhaust pipe – the energy mix used to generate the hydrogen would be approximately 4 per cent wind, lots of coal, some gas and some nuclear – simply the nation's normal electricity mix. Unless the hydrogen scheme buys green power, that glass at the tail pipe should actually contain a few plutonium pellets dropped to fizz away like headache tablets and a gas balloon full of equivalent power station emissions.

Hydrogen is not a primary fuel; it is an energy carrier. It is not a source of energy, but rather a potentially useful way of storing and moving energy around conveniently. It could be useful because in principle a vast over-capacity of renewable energy in some areas could spill energy into hydrogen production facilities. They in turn would supply transport fuel for fuel cell car fleets or pipe it just like natural gas into cities for cooking and heat. Alternatively transport fuel suppliers could buy green power to generate hydrogen.

However, hydrogen can also be stripped from hydrocarbon chains (oil) or obtained from gasified coal. Indeed several coal companies now have their eye on a hydrogen economy. Yet hydrogen produced from these fossil fuels would still produce ample GHG emissions, failing to realize the reductions that are the hallmark of the renewables–hydrogen pairing.

In fact, the thrust of this argument is not to dismiss the hydrogen industry. On the contrary it is to open eyes to the flaws inherent in an approach that fails to think through the end objectives and to measure supported technologies against these objectives. Although in principle it makes sense to apply an even hand or to avoid picking winners, one must take at least some steps to exclude technology that will defeat major social and environmental objectives, that is technology 'dead-ends'.

Killing off future industries through excessive competition

This comes back to the question of objectives. Is the objective to achieve lower-emissions generation, or to develop a market for an entirely new renewable technology? Is the issue cost or industry development? These are potentially quite different objectives and require different policies, and it is quite possible for an over-concentration on emissions or cost to actually work entirely against the objective of industry development.

Lines can become unwittingly blurred when a policy for industry development uses excessive competition. This competition can exist in two ways: first, between technologies and/or, second, between projects. The former is a type of technology neutrality competition, wherein a scheme provides the same support for all renewables on a competitive basis, so biomass, wind and solar thermal electric projects compete equally for the same pool of support. Of course the cheapest projects take up the support and these projects will naturally be dominated by the most mature technology and/or abundant resource. But such monopolization by a particular technology may not be consistent with long-term objectives better served by a wider spread of industry development. The latter point concerns excessive competition between projects, which can result in something similar to a price war. Project proponents reduce their margins to the point where project viability is threatened and successful projects may not actually get built. Furthermore, if the margins are low for the project developers, then they will also be small for all of the suppliers into that project, and history tells us that this does not provide the 'comfort room' that manufacturers in the industry look for.

One of the most impressive achievements of the UK NFFO system was that it did address which industries it wanted to build up and did not lump them together. Having made this decision, it was then a much simpler matter of looking at the support levels required and appropriate market share.

The main scenario that might justify technology neutrality would be one in which there was no interest in building up a domestic industry, only an interest in installing quantities of renewables for other reasons. Even in this scenario this policy would have quite a short-term view, as it would only reflect the pricing structures for renewables built up internationally, rather than addressing the long-term resources available indigenously. This point is raised in Chapter 9 on Spain, a country which possesses a massive solar resource and is demonstrating leadership in solar thermal electric research and development. However, as the following excerpt from this book's case study on Spain shows, that country has not taken full advantage of these assets:

> *at the end of the 1970s the Plataforma Solar de Almería was created as a technology R&D centre and three different, small solar power stations were built, with mirror modules that concentrated solar radiation to levels sufficient to generate electricity in a thermodynamic cycle.*
>
> *The recent bonus approval, equivalent to €0.18 (US$0.24) per kilowatt hour mentioned previously makes proposals for four more 10–50MW solar thermal electric plants possible... However, what is less than ideal is that the bonus scheme for this type of technology was not introduced alongside other renewables at the outset in 1994/98, but had to wait four years.*

Table 2.1 *Total numbers and capacity of projects given NFFO,*
NI-NFFO and SRO contracts, by technology

	Number	Capacity (MW DNC)
Biomass	32	256.0
Hydro	146	95.4
Landfill gas	329	699.7
Municipal and industrial waste	90	1398.2
Sewage gas	31	33.9
Wave	3	2.0
Wind	302	1153.7

Note: DNC: declared net capacity (see Chapter 6 for more information); NI-NFFO: Northern Ireland Non-Fossil Fuel Obligation; SRO: Scottish Renewables Obligation.

Pitfall: Incorrectly targeted measures

A classic pitfall with capital-intensive technologies like renewables is to focus support on reducing the capital burden while failing to require a strong performance from the installations. In the US and India early support schemes resulted in poorly located or operated wind turbines through an excessive use of tax breaks and this allowed poor installation and performance standards to take root.

This classic pitfall has been discussed many times over and is responsible for several skeletons in the closet of the renewable sector. A friend, returning from the US a few years ago, made this comment: 'Oh yes, I've seen windmills. We were driving through California and there they were, huge machines, very impressive. And boy the wind really was howling through there. Is there any reason half of them weren't turning?'

There is a difference between installing renewable capacity and getting renewably generated energy to the consumer. This feat entails more than just an 'on' switch. The crucial step is having all or a major part of the incentive mix attached to the actual production of energy.

Renewable energy sources themselves are often available at low or zero cost, and the bulk of project expenditure lies in the capital cost and installation. These high upfront costs deliver a heavy interest repayment burden on any project, thus incentives that reduce this burden can be very attractive. Capital-oriented fiscal incentives that have been used include investment tax credits, tax holidays, accelerated depreciation, low-cost loans, and direct subsidy.

What is missing in these incentives is a pressure to increase performance. Assuming the capital outlays are fixed, developers can increase their revenue from this investment in many ways, including: improved siting to harness more resources; improved management systems to increase availability and reduce outage; and operation, maintenance and upgrade programmes to maximize the project life.

A balance must be ensured between reducing the commodity price through capital-oriented fiscal incentives, and longer-term incentives to force the industry towards higher performance standards. India in the mid-1990s had to respond sharply to an apparent abuse of tax incentives, and now has a more sophisticated incentive framework as Bakshi points out in this book's India case study:

> *Incentives available for setting up wind power plants in India include: a 100 per cent depreciation in income tax assessment on investment on the capital equipment in the first year; a five-year tax holiday on income from sale of generation; industry status, including capital subsidies in certain states; banking and wheeling (moving) facilities; buy-back of power generation by state electricity boards at remunerative price; and third-party direct sale of power generation in certain states. These incentives coupled with the steep increase in demand for energy has led to an unprecedented enthusiasm by private companies to set up wind farms.*

Pitfall: Opaque incentives

The renewable energy industry is comparatively small, but it is global. Opening up a market in a country new to some of the renewable technologies will require the involvement of non-indigenous companies. To attract the attention and trust of these companies, the policies intended to draw them must be as clear and open as possible.

I write this chapter on a computer that also contains emails from very large companies in Germany, Spain and India asking to be updated on the Australian market and its federal Mandatory Renewable Energy Target (MRET) legislation. Some of these companies are funding industry association campaigns in Australia even before they decide to open an office here. Why? Because they want to be clear about the legislative environment into which they will move before they take the risk.

Scratch the surface of the global renewable energy industry and it is immediately apparent that each technology has a strong geographical focus – wind power in the EU, solar power in Japan, small hydro in China, geothermal in Indonesia and so on. Almost by definition a renewable energy industry entering a *new* market is coming in some part from overseas.

With technology choices attuned to local resources, one can expect some local knowledge. However, this knowledge must be coupled with international expertise if the wheel is not to be reinvented in every new market. If one considers the growth rate of renewable industries, which climb above 30 per cent per annum in some years, one can see that experience is in short supply globally. Companies expanding their operations will therefore direct their focus to markets already developing, rather than speculate on potential markets. Thus clear incentives must be in place to attract the attention of renewable energy industries, or more specifically the companies they comprise.

This need is not always matched by the backroom negotiations that often precede large infrastructure contracts. Governments and companies will negotiate the best terms they can, sometimes with undisclosed incentives on the table. 'We'll give you a tax break this big if you put your factory there,' or 'We'll promise this many jobs, if you guarantee this price for so many years,' and so on. That is not the kind of incentive programme that will attract a foreign company to set up an office. Rather it will be perceived as an insiders' market.

The more obscure and buried the incentives in the idiosyncrasies of host country policy, the worse it becomes for foreign companies. I once heard a conference speaker explain how they had just managed to get one of Italy's first big wind farm off the ground despite legislated incentives being in place for two years. They undertook a joint venture with an international legal firm which had to spend over six months dissecting the legislation to determine how to earn some profits from a wind farm.

Contrast this with the German feed-in law that effectively states that anyone who makes renewable energy will get a fixed percentage of the residential sale price per kilowatt hour (kWh). Full stop. This may not be the best industry development mechanism, but there is no disputing its clarity, which acted like a beacon to investment and industry for well over a decade. The more complexity, the more caps, strings and sunsets, then the more hesitant industry and investors will be to get involved.

Pitfall: Boom and bust – a lack of policy and market stability

Just as simplicity and clarity can act as a beacon, so can stability. Conversely, lack of stability severely increases risk, risk costs money and that makes renewables less competitive. Even apparently stable policies can lead to a boom–bust cycle through the use of inappropriate caps and sunsets.

Industry rarely trusts government goodwill; it is not bankable. Instead industry looks to policies and legislation, because the contracts that flow from such legislation are bankable. However, three things can undermine confidence in those policies: volume caps, time frames (sunsets) and excessive policy change.

Caps and targets

What starts out as a target must by definition become a cap if that target is successfully reached. Caps are great for limiting problems, but are a nightmare for building solutions. Experience indicates they lock in a boom–bust cycle.

Renewable projects often have a long lifetime (20 years or more), but all of the investment and the main industrial activity occurs at the beginning. In order to maximize the returns on an investment, developers will try to run the project for as long as possible. Thus, to meet a production level, they will try to establish capacity as early as possible in the scheme to yield the maximum return time. Generation schemes with targets/caps that run for less than 20 years will produce a flurry of activity and massive industry growth for the first few years of the scheme. Once capacity meeting the long-term energy cap is in place, a complete halt in activity will ensue.

Even longer-term legislation can fuel a boom–bust cycle. The Australian MRET legislation, which has a ramped target to 2010 and a sustained level to 2020, has resulted in a boom–bust effect which is already well underway. The

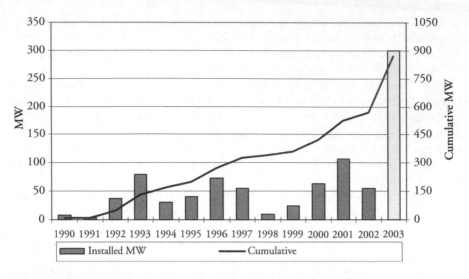

Note: Data for 2003 are forecast
Source: BTM (2003)

Figure 2.7 *The boom–bust cycle has occurred in an extreme form in the UK due to the multiple use of limited targets*

original target was for a 2 per cent increase in renewable energy generation in the power sector from a 1997 baseline. Given growing energy demand, even this modest target would have had an added ongoing increase; however it was revised to a fixed target of 9500 gigawatt hours (GWh) to provide more certainty. In fact this will mean a renewable market increase of less than 1 per cent by 2010 and a diminishing contribution from renewables as wider energy demand continues to be met by coal.

The renewables eligible under the MRET were very active in the first three years of the scheme, but the end is already in sight. For example, the estimated capacity required from wind is about 1000MW. Installed capacity growth averaged nearly 120 per cent annually in the three years following 1999. The installed wind power capacity at the end of 2004 was 380MW, but more than 5800MW of projects have been identified by developers and are at some stage of development, while about half of these had applied for planning approval in mid-2004.

Only the fastest to secure contracts and install will succeed, the rest will be left out. Almost all of the projects installed will be in place within the first five years of the scheme. Thereafter the Australian wind industry hits a wall.

A boom–bust cycle is a terribly inefficient form of industry development and is the complete antithesis of what is required to establish manufacturing. If targets are used they must be dynamic to provide a constant but steady pull on industry.

Sunsets

The second is the problem of sunsets. (This particular business expression makes me think of the sun dropping into the ocean at the end of wonderful day.) It generally refers to how a government, investor or industry winds down a scheme, namely, 'When does it finish?' Perhaps to keep the treasurer happy or to put a fixed cost on an initiative, governments often like to build in a sunset.

This is another example of cross-cutting objectives, where the desire for economic certainty overrides the objectives of industry development or climate mitigation. Given that effective climate mitigation will require a 50–100 year transition to zero-emissions, safe energy provision, what on earth is the point of a scheme that runs to 2010, or 2015? Worse still, a sunset tells investors, 'Don't get too comfortable because we are not in this for the long term.' It may provide the treasurer with apparent certainty of the scheme's economic impact, but this is a mistaken certainty since the prices of renewables and conventional energy change all the time.

One way to achieve exactly the same outcome, but with the ability to provide the basis for ongoing commitments, is to replace sunsets with reviews. The assumption must be one of continuance and stability, subject to the performance of the industry and the evolution of external factors.

The problem here is that, for government, one parliamentary term is about four years, two policy terms is guesswork, 10 years is a long time and 20 years is somebody else's problem. I have sat in meetings with very senior national bureaucrats, responsible for long-term energy planning, who have told me with great gravitas that they are looking 20 years ahead. Twenty years is less than the lifetime of most renewable installations, from a PV panel to a wind farm and it is half the life of many thermal power stations – hardly an adequate time frame to achieve a fundamental sectoral transition.

Excessive policy change

As one does with an associate who keeps breaking commitments or changing arrangements, industries too will lose confidence in government if legislation keeps changing. They may depart to do business with more reliable governments or expect higher returns to cover the risk of doing business in such an environment. Alternatively they may simply go bust! The following excerpt from this book's UK case study provides illumination:

> *Also detrimental to the development of a thriving renewable business sector in the UK was the stop–go nature of NFFO. Bidding rounds were irregularly spaced, with relatively little indication of when the next round might occur. It is notable that a review undertaken after the second round left a three-year gap between tenders being invited, during which time many of the fledgling businesses collapsed through lack of business, or left the industry due to dissatisfaction with the process.*

Pitfall: Inadequacy/excessive fiscal constraint

Deep at the heart of any treasury department lies the horror of spending a cent more than necessary. Transferring tax dollars to the private sector is a flow in the wrong direction for any well-meaning treasurer. And so a pressure for 'economic efficiency' emerges.

The argument was put to me recently by a government bureaucrat in terms he referred to as 'financial discipline'. I had just run through some projected cost curves for wind power to 2020 and compared these to the cost of fossil fuels and some likely costs of carbon. My point was that convergence in pricing was not some nirvana off in the haze, but something potentially achievable within 20 years, less than the life cycle of the current projects. 'In that case,' he asked, 'why are you demanding high prices for renewable generation? Why can't we force the industry to stick to these low prices and keep our schemes as cheap as possible?'

The answer is obvious if you look though the eyes of the industry that we are trying to attract and motivate. 'The set-up costs are high and the risks are higher still. Even though I want to diversify my activity base, I've got more than enough business on my hands. So, if I'm going to set up somewhere new I need to be sure that there will be enough margin and enough of a future market to ensure I don't come out having wasted my money or wasted my time.'

New industry development policies need to work like an automobile choke, which is the knob one (used to) pull to make the fuel–air mix of the carburettor rich enough to start the car's engine. Once the engine is warmed up and running smoothly, the choke is pushed back allowing a lean fuel–air mixture to keep the engine running optimally. Starting a cold industrial engine without making the financial mixture rich at the beginning is a waste of time and simply results in the battery (the start-up support) being run down. The quicker the engine starts, the quicker it gets warm, the quicker the fuel mix can be made more lean, and the quicker the battery is replenished by the running engine.

There is much discussion in the renewable industry about mechanisms. However, the above principle is not dependent on the mechanism, fixed price or fixed volume, as these can both be structured to create a strong sellers' market. The proof is in the pudding if we compare the UK NFFO approach, which kept prices as low as possible and left maximum risk and uncertainty, with the German approach that was much more lucrative and stable and left room for movement by industry and investors. Germany and its consumers may have spent more in the short term, but now they have 120,000 employed in the renewables industry contributing to the country's tax base. I would argue that this approach is more fiscally efficient!

Oversights

Oversight: Absence of contextual frameworks

Unfortunately this 'contextual' oversight seems to be an almost textbook move in new markets. In this instance, a renewable energy driver is established and then renewables are thrown into a system which lacks knowledge and experience on handling the newcomer within existing frameworks. Expecting renewables to proceed in the absence of contextual frameworks is asking for trouble.

In a new market, it is virtually guaranteed that all the regulation into which renewables must fit will not have been designed with renewable energy in mind. For example, a licence to generate electricity may reflect the cost of a half-billion-dollar power plant. Will a 10MW run-of-river hydro scheme have to purchase that same licence? Or more pointedly still, will someone who puts 100 watts of solar PV on their roof?

Whose approval is needed to install a machine? Whose approval is needed for connection? Is the fuel certified or standardized? Who expects payment? Is a renewable power project a property development, an industrial installation, a power station or a piece of farm machinery? How much should be charged in rates? Is a solar panel a power station or a home appliance? What do we do if the meter goes backwards?

At a recent talk, wind energy writer Paul Gipe described an instance in which wind turbines were classed as power stations and required to withstand an earthquake without catastrophic failure. It is one thing to require that a power station on ground level not topple, but 50 tonnes of steel on top of an 80-metre tower – that's a big task! The consequences of an unmanned tower toppling in a farmer's field cannot be compared to the flattening of a thermal power station full of workers.

Thus contextual frameworks must be checked for all the legal processes that renewable projects will need to pass through, from their inception to decommissioning. Stepping through this process reveals the many points at which renewable energy projects interact with the outside world. With care, one should be able to spot almost all of the contextual issues that will arise in a new market.

The next trick is to have the mandate to deal with them. Some areas to look out for include project licensing, rates, building regulation, zoning and planning, licences to generate, grid access requirements, generation standards, liability and insurance, charges or remuneration for transmission of power and environmental impact.

Oversight: Energy market reform and access

Electricity grids were hitherto developed in conjunction with large power stations and extensive distribution systems. The policies that regulate these grids and the power markets that use them require major reforms to render them equally applicable to renewable and embedded sources. Oversight occurs when renewables are expected to work within a policy system created at a time when nobody imagined electricity flowing any way other than from inside the system and out.

Today even major interests are suggesting that new generation may be small and distributed. Electricity will be more like an interlaced canal system, with some putting water in and others taking it out – hence the popular term 'electricity pool'. We may come to see small gas co-generation systems in the bottom of virtually every big building firing up to provide peaking power. We may see wind farms at the end of grids and suburbs of solar roofs pumping out power to commercial districts during the day. Unfortunately the 1950s model of large plants and lots of

Box 2.4 *When policy fails: Guerrilla solar*

One alternative to policy reform is guerrilla warfare. At the European PV conference in Glasgow I was introduced to a group of American engineers from what is now called Schott. (They make sine wave inverters for solar panels and sport T-shirts that read, 'We invert 'till it hertz.' I gave this T-shirt to my partner, who teaches yoga and spends preposterously long periods of time upside down).

It was a bright and calm spring morning. Two shadowy figures emerged from an unmarked van. It was so plain, so ordinary, that if anyone had been around, they would not have noticed the activity on the otherwise deserted city street.

One person was carrying a long case with a handle. The form quietly stepped across the sidewalk and glanced in both directions. In a graceful, practiced manner, a key was eased into the lock of the building. The figure slipped inside so quickly that it seemed as if no one had even been there.

The other person moved cat-like, gently closing the side door of the van. In through the unlocked door of the building, the second was soon behind the first— carrying a thin but somewhat unwieldy cardboard box about four feet square. It was carried as if it were precious cargo, like an expensive 17th century painting by one of the masters. The door clicked shut, and locked.

They were pros. This was guerrilla solar.

Chapter One
Earlier this year, we made a toll-free phone call to Alternative Energy Engineering (AEE) in Redway, California. A new solar product was on the market, and we were excited to check it out. We ordered the new "AC Module." This 108 Watt 24 Volt PV panel has Trace Engineering's Micro Sine inverter glued to the back. The inverter box on the module back is not much larger than the junction box itself. The inverter is pre-wired into the module's junction box, and has a long, thin, four-wire cable coming out of the side. A standard wall plug is wired onto the end of two of the wires, and then inserted into an ac receptacle inside the building.

Our intentions are honorable, but less than legal. We did not tell the utility that we were going to feed homemade electricity back into their power grid. This wasn't a decision arrived at lightly. Any revolutionary action should have all of the potential consequences weighed carefully before proceeding. In this case, we knew the risks would be minimal—unlike blockading a nuclear power plant where arrest was likely. We felt that the worst scenario would be the utility shutting off our power until we de-installed the illegal PV system.

Safety Isn't the Issue
Every utility puts high priority on making things safe for

Source: Home Power (1998)

Figure 2.8 *Guerrilla solar in action*

These people introduced me to the shadowy world of 'Guerrilla Solar', bands of wanton criminal types who wear balaclavas to disguise their true identity whilst they put solar panels on their roofs and *plug them in*. I am informed that in many US states this very act violates a raft of laws designed to keep good citizens from doing anything but consuming electricity. No licences for power production, no certification, electronic devices sending signals into the grid – all serious offences. To top it all, these audacious guerrillas take photographs of themselves in the act and post them on the web.

In fact, my new friends considered themselves a cut above this approach and instead were considering a strategy of re-branding. Their masterstroke was to turn the plugging-in concept on its head. Solar panel micro-sine inverters hitting the market at the time allowed one to essentially plug the solar panel into a household socket. There were, however, many grey areas about exactly how to classify a solar panel and what rules it would need to comply with. So these lateral thinkers were ready to classify their solar panels as Christmas tree lights. Every year hundreds of houses across the US go up in flames because Christmas tree lights have so few standards to meet. By putting two light-emitting diodes (LEDs) in the line from the panel to the plug product would be both the safest and most energy-efficient Christmas tree lights in the US.

wires does not fit this evolution and it must be changed to allow renewables and distributed sources to enter the game.

Energy market reform and access oversights vary hugely between electricity systems. But certain common problems seem to crop up time and time again. For example, the owner of a power network may have the ability to refuse network access. Or that owner may be able to apply excessive or unreasonable costs, or gain windfall payments through the auctioning of spare line capacity. On a different note, the infrastructure planning system is unlikely to have ways to factor in savings though embedded generation and reduced system use (least-cost planning). And a different distortion can occur when the local value of energy (i.e. the cost including generation, distribution and losses) is not transparent or disclosed. Finally, the licensing for generation will be likely to be based on large power stations and may be inappropriate for small-scale generation.

On a more long-term policy front, similar issues must be confronted in terms of how grid infrastructure is extended, reinforced and paid for. How are licences calculated and assigned? How are power quality conditions specified? How are forward contracts for power set up and over what time periods? Are conditions the same for all power providers?

Oversight: Poor risk/cost–benefit distribution

Any excessive burden without equivalent benefit for a given stakeholder (whether local, regional or national) is likely to result in some sort of opposition to renewable energy developments. It is a dangerous and potentially irreversible policy oversight to assume that this balance will occur by itself or assume it to be outside the role of policy-makers.

Box 2.5 *The embedded generator and the distributor's killer move*

Imagine a small run-of-river hydroelectric scheme, located at the end of a long grid line that requires lots of upkeep and wastes ample energy through line losses. The hydro developer has a power purchase contract with a utility in the city, but the developer plans to take advantage of the savings she can offer to the local utility to make up the project income. 'I can save the local distribution company from having to upgrade that old line and also reduce their line losses. If they give me a portion of these avoided costs my project becomes viable!'

But the local distribution company has different ideas. 'You're producing electricity here and selling it in the city. That means you need to pay us to get it there for you,' they argue.

'Hang on,' says the developer. 'In principle that's true, but in practice I'm just reducing the amount of electricity you need to pull down that line, which reduces your service costs to the transmission operator. You should pay me.'

Then the local distributor makes his killer move. 'I decide who gets connected and on what terms. Take it or leave it.'

In practice this means that the benefits of renewable energy accrue at different levels. Pollution mitigation occurs at a global level, inward investment occurs at a national level, manufacturing and employment creation can occur at a regional level and some employment will occur at the very local level. However, the direct

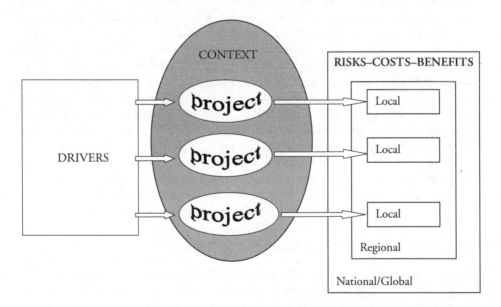

Figure 2.9 *The drivers and context will largely define the risks, costs and benefits to local, regional and national stakeholders*

physical impacts occur at a local level only. If the actual or even perceived impacts are significant local people may object to a project. We will consider this issue in more detail below with respect to planning.

There may be many ways to minimize or reassure local people about what the actual impact of a project will be. However, there is also the avenue of increasing or targeting additional benefits to help achieve a desirable balance for a given set of stakeholders.

Experience in Denmark indicates that there is a strong link between local social acceptance of renewable developments and local ownership. As we will elaborate in the next chapter, the highest concentration of wind turbines in the world occurs in a place called Sydthy, Denmark. The reason that the community accepts and allows so much wind development may be explained by the fact that 58 per cent of the households in Sydthy have one or more shares in cooperatively owned wind turbines.

This use of the benefit of equity ownership (encouraged in Denmark by favourable government tax policies targeted at residents local to projects) balances risks in a way that leads to much greater acceptance of projects and their impacts.

Oversight: Absence of commensurate planning and planning reform

Land use planning is another critical area of commensurate policy-making crucial to the multiple, small-scale, geographically distributed projects that typify renewable energy. To install the same capacity as a thermal power station, 10 renewable projects may be required; to generate the same energy, perhaps 30 projects. Given these numbers, excessive constraints on planning consent may constitute an oversight which can bring the industry to a grinding halt. Planning is another example of where the renewable industries enter otherwise uncharted territory to encounter policy voids.

The UK provides a cautionary tale of how the inertia in the power market can leave developers and stakeholders without a commensurate planning framework. Despite possessing one of the highest wind resources in the EU and one of the earliest promotion policies legislated in Europe, only 18 of 55 awarded contracts in the UK received planning consent, according to Lloyd-Bessen (1999). The following excerpt from Edge's case study brings this point home:

> *While the structure of NFFO made life difficult for developers, more capacity would have been built if there had been stronger guidance from central government to local planning authorities. In the absence of this steerage, councils were often swayed by small, but vociferous minorities who objected violently to virtually any application... A string of planning decisions followed which turned down applications for wind farms on the grounds that the landscape damage that would result was not justified by the small amount of power that would be generated. This was inconsistent with stated government policy, but without this policy being translated into direction in the form of a Planning Policy Guidance note, local officials were free to ignore it.*

The complex NFFO bidding process was only really accessible to professional developers and financiers, so local communities in the UK were effectively shut out – hardly a situation likely to breed trust and mutual support.

This situation contrasts strongly with the Danish experience, as described by Krohn (2002):

> *With a highly visible technology such as wind turbines, the development of models for dealing with public planning (zoning) issues have been very important for many countries' acceptance of the technology. In Denmark the public planning procedures were initially developed though local trial and error. In 1992 more systematic planning procedures were developed at the national level, with directives for local planners. In addition, an executive order from the Minister of Environment and Energy ordered municipalities to find suitable sites for wind turbine siting throughout the country. This 'Prior Planning' with public hearings in advance of any actual applications for siting of turbines helped the public acceptance of subsequent siting of wind turbines considerably. A similar planning model has since been introduced in Germany with considerable success. Other countries are studying these experiences with a view to overhauling their planning procedures.*

Thus we see that the ease and simplicity of obtaining planning or installation consent becomes crucial to this development-intensive industry. Overlooking this issue can stall the industry – or worse.

Conclusions

In this chapter we put the dirty washing out for all to see. The aim has been to ensure policy failure is understood so that we can be sure we come up with a policy that works in the next chapter.

The first stumbling blocks we noted were myths which can misinform decision-makers about renewable energy before they even consider policy-making. One major myth concerns technology and assumes the price of renewable energy will be reduced by technical breakthroughs. There is also the myth of the righteous, which assumes a good idea will always succeed and that intervention is unnecessary. The hands-off myth erroneously assumes that government intervention undermines the proper working of markets. And finally there are the money myths, which assume that renewable energy is more expensive than thermal power, and that because it is expensive and capital intensive the best governments can do is to throw money at the industry.

We also examined the pitfalls waiting to trap unsuspecting policy-makers even if they are well informed and have solid intention. These include under-defined objectives, inadequate resource or technology identification, incorrectly targeted measures, opaque incentives, inadvertently setting up boom and bust cycles, as well as a lack of policy and market stability, and the counter-productive nature of excessive fiscal constraint.

Finally we saw the results of supporting renewable energy with the right hand but ignoring what was being done with the left. Here we looked at indirect areas of policy that can cause renewable energy development to become unstuck. These include an absence of contextual frameworks, and an absence of energy market reform and access. Another major issue is poor risk/cost–benefit distribution across the host society. We also saw how the absence of commensurate planning and planning reform could pose a risk to renewable energy development.

Thus, in this chapter, we drew some salutary lessons from past experience. Our next step will be to look at the different bases that must be covered to avoid the same policy mistakes or oversights in a new or expanding renewable energy market.

References

BTM (2003) *World Wind Energy Market Update,* BTM Consult industry analysis report, EWEA, Brussels

Dambourg, S. and Krohn, S. (1998) *Public Attitudes Towards Wind Power,* Internal publication, Danish Wind Industry Association (www.windpower.dk)

European Photovoltaic Industries Association (EPIA) (2003) *Solar Generation,* European Photovoltaic Industries Association and Greenpeace

EPIA (2005) *Solar Generation,* European Photovoltaic Industries Association and Greenpeace International

European Wind Energy Association (EWEA) (2004) *Wind Force 12,* European Wind Energy Association and Greenpeace International, Brussels

Home Power (1998) 'Guerrilla solar', first page of article by M. Rukus and J. Freely in *Home Power,* no 67, October–November

KPMG (1999) 'Solar energy: From perennial promise to competitive alternative', Report 2562, KPMG Netherlands, August

Krohn, S. (2002) *Wind Energy Policy in Denmark: 25 years of Success – What Now?* Danish Wind Industry Association, Copenhagen

Lloyd-Besson, I. (1999) *Lessons from the Frontline – the British Experience of Opposition to Windfarms,* Australian Wind Energy Conference, Newcastle, Australia

Transition Institute (2004) *Cost Convergence of Wind Power and Conventional Generation in Australia,* eds K. Mallon and J. Reardon, Transition Institute Report to the Australian Wind Energy Association (AusWEA), Melbourne

Windpower Monthly (2004) Chart originally printed in *Windpower Monthly News Magazine,* vol 20, no 1

World Bank (2004) *Extractive Industries Review,* World Bank Group, www.eireview.org/html/EIR FinalReport.html

3

Ten Features of Successful Renewable Markets

Karl Mallon

In the last chapter we examined some policy errors, from which there are many lessons to be learned. In this chapter we will ask what action on the part of government is required to get rapid and sustainable renewable energy implementation underway. Taking the need for rapid implementation of renewable energy as a given, we will examine key factors required for policy design and implementation. This chapter is designed to help define a set of policies that avoids pitfalls and provides a solid framework for renewables.

Considering the failures described in the last chapter, we can summarize ten key features of successful renewable energy policy:

1   transparency;
2   well-defined objectives;
3   well-defined resources and technologies;
4   appropriately applied incentives;
5   adequacy;
6   stability;
7   contextual frameworks;
8   energy market reform;
9   land use planning reform; and
10 equalizing the community risk and cost–benefit distribution.

We can classify the ten key features within the framework of drivers, contexts and society. Features one to six are driver-based issues, while seven to nine are contextual issues. The final key feature, equalization, is a societal issue.

It is immediately apparent – and important to recognize – that renewable energy policy consists not just of a driver, but rather comprises a complete framework. Experience in country after country shows that overlooking or ignoring parts of that framework will undermine the entire vision.

Many of the above key features have become mantras for renewable energy lobbyists and will be quite familiar to readers. However, here we have the opportunity to be more specific about what we actually mean when we say 'stability' or 'market reform' and to clearly describe how each component contributes to defining the path through the minefield of potential policy failure.

In each section of this chapter I will pose key questions and then attempt to answer them as comprehensively as possible. However, it is important to recognize that the list of answers may be incomplete and that every national circumstance must be considered according to its own characteristics.

Transparency

In order to kick-start renewable energy markets, support schemes and policy frameworks must be visible and accessible. In any new market most potential entrants will be outside the market at the outset. Entrants may come from energy companies, the civil engineering sector, heavy industry, the finance sector or they may be entrepreneurs looking at niche opportunities. They may also be looking at the market from overseas. What they all have in common is that they are coming from somewhere else.

It is important, therefore, first, that incentives are not created on a case-by-case basis and, second, that incentives are not made available for individual companies to negotiate or to leverage additional incentive. Instead, support must be transparent, accessible and open to all players. A lack of transparency tends to favour insiders or more influential companies and creates market advantages that will deter new players.

So how might a transparent policy look? How must it appear in order to attract the various players needed to get renewable projects up and running?

Some considerations examined here will overlap with other policy objectives on our key features list, but we address them here specifically in the context of ensuring that new entrants are not dissuaded from entering the market because of a lack of clarity and transparency. Thus, when forming or reviewing a new policy, the following questions and answers will give some clues as to its transparency.

Are the policies comprehensible enough to understand and comprehensive enough to cover all of the components required to make projects bankable?

At the outset, we must consider whether the policies we are making are clear and straightforward, and that they cover all necessary bases. We can do this in two ways. We can think of all the possible areas where we may need to legislate, or we can go to a renewable energy project developer and ask, 'What is it that you need to make your project bankable?'

Bankability is the litmus test of any energy project's viability. It means that the project proponent can produce a set of contracts, approvals, surveys and cash flow projections, and set these down in front of a bank or financier. If the bank sees all the contracts in place to allow the project to proceed and deems the projection for the revenues to be generated to be reliable and adequate, then the bank will typically be prepared to lend say 70 per cent of the project's value. If any components are missing, banks will not fund the project. If some components are uncertain, this may translate into risk premiums that increase the cost of the finance.

Such benchmarks are set by the financial sector, not government. So it is to the government's advantage to ensure its policies cover all such bankability issues, and to do it in a manner that makes it clear to new and potential entrants that this has been accomplished. The policy components required are fairly standard (varying between technologies to some degree). These issues include rights related to the location (land or building), market access, taxes and charges, fuel costs, the value of produced energy and so forth.

The key message is that these components are essential to the banks and therefore the industry, and must be clearly and transparently presented within the renewable energy policy framework.

What unknowns exist in the policies that might affect the size of the market, the prices paid for renewable energy or the duration of the scheme?

Unknowns are like the dark matter of renewable energy; they can mean the difference between the universe exploding or collapsing. Some of the variables that can interfere with market growth include: the level of support, the duration of support eligibility, the duration of support schemes and the quantity of renewables required under the scheme (upper limits). Again, however, it comes back to increasing the bankability of projects and minimizing the uncertainties that expose projects to risk.

Many would agree that one of the factors responsible for successful German renewable energy policies is the minimization of such uncertainties. The price for various types of energy was clearly set out and the scheme did not have cut-off dates within the lifetime of likely projects. Developers still had to deal with variables including hardware costs, fuel costs or size, exchange rates and so forth, but thereafter the system into which they were selling was to a large extent set in stone (subject of course to changes in legislation).

In market-based systems, a renewable energy commodity price introduces a new variable and a new risk. Risk increases costs, but market dynamics can also reduce prices. So there is a balance to be struck and a policy choice to be made. How stable is the market? How far out into the future is it defined? Is it likely to increase or decrease in size? These are all very relevant uncertainties that must be removed or minimized if possible through policy.

Manufacturing is particularly susceptible to unknowns. Developers can always come into a system, put up some projects and leave again if things prove too hard or, if things are risky, they adjust their prices accordingly. However, hardware manufacturers require that a successful market work for several years before they cover their costs. For them risk and uncertainty is not a price adjustment, but a threshold. If the risk is too high it is better to import at a higher cost than to expose the company to the risk of million-dollar factories lying idle.

Are the policies structured fairly so that they do not favour insiders compared to outside entrants?

In practice, the policies put in place should be as universal as possible. They must be open to the widest possible base of businesses, kept free of time constraints or cut-off dates,[1] and based on standard requirements that are transparent or not subject to interpretation.

Recently an Australian government agency issued a call for new bio-fuel manufacturing projects for which it would provide capital grants. The deadline for applications was set within a couple of months of the issue date and there was a fixed amount of money to be spent. Applications required full business plans and verifiable supply chains.

This is an example of a policy that not only favours incumbent players, but is also anti-competitive since it penalizes anyone wishing to enter the market later. Latecomers not only miss out on assistance, but have to compete with the successful applicants who by then have a subsidy. Policies that are not equally accessible to any renewable energy business will actively dissuade new entrants and at their worst can appear nepotistic or politically expedient.

Do the policies have sufficient time frames to allow dissemination and engagement from interested parties in other countries and sectors?

Short-term schemes are like a gold rush. They create cities overnight and leave behind ghost towns.

The best outcome for a government occurs if the maximum number of businesses is attracted to participate in a new industry. This will translate into high production of renewable energy and competition to drive down prices and accelerate industry development.

As we have mentioned, many entrants will either be coming from other domestic sectors or from overseas. In both cases, they need time to learn policies and measures, to investigate the market and/or country, to establish themselves in the industry and hire employees and offices, and thereafter to start doing business. This all takes several years. Given that projects last for 20 years and need price certainty for the majority of that period, the policy framework requires a run time which allows for this crucial set-up period as well as a reasonable period of doing profitable business.

To put some sort of number on this time frame we can note that most renewable energy installations will have a lifetime of at least 20 years. To keep the prices for produced power as low as possible, eligibility for support should be at least 15 years. Otherwise developers must increase prices to recover the investment. Therefore a policy scheme that lasted no more than 20 years would foster high activity for five years and then result in a standstill as far as new activity is concerned. A scheme with no fixed term, but instead with a guarantee that all projects are eligible for at least 15 years of fixed support from production start-up, provides continuous and stable installations for as long as the scheme is in place.

Are the renewable energy policies consistent with policies and measures for other parts of the energy sector, or are there mixed messages and double standards?

We will discuss the consistency of policies below, but we must also raise the issue of the consistency of messages. We need to consider specifically the clarity and consistency of the messages going out to potential market entrants.

Government policy, or rhetoric, sometimes from various departments, is not always fully aligned; rather it is dynamic and may be contradictory. This affects renewable energy investment when there is inherent conflict between statements and policies on energy or the environment.

Any business must acknowledge that legislation can change from term to term. In so doing, business must look to longer-term indicators to assess possible futures. Thus a business considering renewable energy investment will be looking closely at government policies towards climate change, energy pricing, pollution or extended producer liability.[2] I will also be looking at the relative support government gives to other industries. If a government spends millions of dollars on fossil fuel research but nothing on renewable energy research, what insight does that provide into its views on the future energy mix?

Well-Defined Objectives

Does the driver deliver the intended outcome? A simple question, but does it actually get asked and are the policies that are devised held up in this light?

The rationale for creating renewable energy drivers varies from country to country. However, in all cases it is important that the policy should be constructed to ensure that objectives are actually achieved. If it is not, implied or expected outcomes not solidly built into the policy may never be delivered in practice. As in the previous section, we will ask and attempt to answer some key questions that relate to this key feature of energy policy.

What outcomes are actually intended from the renewable energy policies?

There are many reasons (and therefore many potential objectives) for accelerating renewable energy development. They include sustainability objectives, energy policy reform, renewable energy production, new generating capacity, indigenous fuel manufacture, greenhouse gases (GHGs) mitigation, distributed generation, increment size, energy cost and least-cost planning (internalization), energy security, new industry/manufacturing development, development of intellectual property in new technologies, job creation, rural development and nuclear phase-out.

From this basket of potential benefits, the next step is to prioritize them for the country in question. For example, the government of a small island developing state may determine that additional electrical generating capacity is its highest priority, while also being concerned about increasing dependence on fuel suppliers and volatile world prices. So the government chooses renewables to lessen external dependence and improve energy security. Furthermore, with its small economy, it is very price conscious and not so concerned about favouring one technology over another. This government wants to avoid mega-projects that create big debts and it also wants low prices for electricity. It is especially pleased if prices fall below those of its current diesel generation. The use of renewables in reducing its GHG emissions and increasing the state's sustainability levels is desirable, but the state has no legal obligations and consequently these are a low priority.

So, in this case, the country's renewable energy policy priorities, in order and beginning with the most important, will be: new generating capacity, reliability, energy security, small increment size, low-cost generation and finally sustainability objectives.

Alternatively a large European country with international GHG emissions reduction obligations might prioritize this way: energy security first, followed by GHG mitigation, job creation, nuclear phase-out and finally the development of intellectual property in new technologies. In either case, the important thing is to recognize and integrate these priorities into policy.

Are the drivers and measures specific about the intended outcomes?

Once the list of intended outcomes is identified, the policy framework must be built to ensure its actual delivery. If significant domestic manufacture of renewable energy hardware is an objective, for example, how explicitly is this set out in the policies? If GHG mitigation targets are intended, are they incorporated as minimum targets in the policy, or set as milestones against which to judge and adjust the policy if it fails to deliver? The following passage shows how Spanish renewable energy development had to reflect a priority for regional economic development:

> *Whilst national laws are important, a crucial impetus for wind development in Spain has come from the bottom up, from regional governments keen to see factories built and local jobs created... the incentive is simple: companies who want to develop the region's wind resources must ensure that the investment they make puts money into the local economy and sources as much of its hardware as possible from local manufacturers.* (EWEA, 2002)

Clearly the message here is that the drivers need to be as specific and explicit as possible about what they actually drive. In some cases it is possible to link the support made available by the driver to its intended outcome. A classic lesson learned here is evident in the failure of schemes which provided incentives for hardware but which failed to link the support to companies actually running the hardware. Wind turbines in the ground but standing idle – this was not intended! We now know that connecting price support to the number of kilowatt hours or litres of renewable energy produced is one way to achieve a desired outcome.

Ideally we try to apply the same principle to all of our intended outcomes. It would be difficult for one driver or mechanism to cover all objectives, but it is relatively easy for a portfolio of measures to do so. For example we may want short-term industry development of solar power, as well as short-term carbon emission abatement. To get both in one policy would be almost impossible. However, it is quite sensible to have carbon trading used in conjunction with a renewable energy support scheme. The former would deliver short-term emissions reductions, while the latter would provide a long-term strategy for convergence between the renewable industry and the wider carbon market as prices come to merge. If domestic manufacture is a high priority, to cite another example, it might make sense to link support to levels of local content, or provide tax breaks or import tariff regimes that give an advantage to locally manufactured or assembled equipment.

The take-home message for this question is that outcomes should not be left to chance, but be made specific and explicit.

Box 3.1 *Revision in Spanish feed-in tariffs to reflect cost changes*

The Spanish renewable energy system is held up as a successful and yet simple scheme that has catapulted Spanish renewable energy companies into being major global players in both the wind and solar sectors. The system used, however, has some built-in flexibilities that allow it to be optimized based on the variations of the market.

> The special regime had been regulated from 1980 through diverse documents. However, arising from Act 54/1997, it became necessary to develop the Royal Decree 2818/1998 to adapt the operation of that regime to the new situation. That situation included new legislation and the introduction of competitiveness on the electric market. With this aim, the producers under the special regime were given guaranteed access to the grid. On another hand, a legal framework was established concerning the sale price of the 'green' electricity produced from renewable energy source. In this context, the Government developed a mixed system that allowed a premium to be applied, within a percentage band linked to the average tariffs, or to a fixed price.

> The producers could opt for the premium or the fixed price, indistinctly, and change their choice every year, according to their corporate strategy. The premiums and fixed prices have been amended every year, in coherence with the market circumstances. (Segurado, 2005)

This system cuts both ways since it ensures that, as the renewable energy prices for a given technology evolve, the scheme does not result in excessive profits to developers. However, it also means that if the market conditions in which renewable energy companies are operating become excessively adverse, the value of the renewable energy premium can be adjusted to maintain profitability.

So the Spanish system, through its very flexibility, provides a great deal of security to industry participants and also maintains the efficiency and integrity of the system on behalf of the consumer.

How is performance checked and the risk of under-performance minimized?

Despite best efforts being paid to the above issues, there may still be areas which defy definition or objectives that remain at odds to some degree. Many renewable policies may be a mix of specification and implication. A policy that provides support for manufacture cannot guarantee that manufacture will occur unless government builds the plant itself. This policy can only make the framework as conducive as possible and be as specific as possible about what the policy is trying to induce. However, the government can check to see whether the policy is working.

While we must consider policies to be dynamic, they should not be unstable. Establishing milestones that reflect objectives, then reviewing policies on a regular basis against these milestones is probably the most effective route.

Indeed policies can be built with adjustable parameters that re-optimize said policies on a regular basis. For example, both Spain and Germany have revised feed-in tariffs to reflect changes in renewable energy production costs. The German system for wind was adapted to allow for price reductions due to technology advances, but also price increases for developers establishing wind projects in less windy (and therefore less profitable) locations. In this case the driver was adapted to clearly incorporate the objective of geographical diversity for wind developments.

If the objectives are well specified and milestones set out, then, accordingly, the respective parameters on support, targets, prices, tax benefits, trade tariffs and so forth can all be adjusted, and new milestones can be established. In this way, the delivery of specified objectives takes primacy and the policies are partially future-proofed without undermining stability.

Well-Defined Resources and Technologies

The need to introduce a new policy for renewables assumes that the existing market conditions will be unable or too slow to deliver a certain set of technologies. However, for the policy to do its job properly, the targeted technology set must be defined in some way. So what do we mean by renewable energy?

The technology sets can either be defined by resource/technology or by outcome, notionally referred to as 'technology specific' or 'technology neutral' respectively. For example, the Japanese may have deemed special support for solar photovoltaics (PV) to be desirable because of its potential to open up a major new market for their electronics industries. On the other hand, portfolio market-based schemes (as described in this book in Chapter 6 on UK policy) are designed around generic industry development of a renewable sector with least cost being a guiding principle.

However, the net for technology neutrality can be cast a long way. We cannot always assume that the policies within which renewables function will exclude non-renewables. The questions and answers discussed in the following sections explore the consequences and risks associated with deferent definitions of eligibility.

Box 3.2 *What is new renewable energy?*

According to the International Energy Agency (IEA, 2002):

> *Renewable Energy is energy that is derived from natural processes that are replenished constantly. In its various forms, it derives directly or indirectly from the sun, or from heat generated deep within the earth. Included in the definition is energy generated from solar, wind, biomass, geothermal, hydropower and ocean resources, and bio-fuels and hydrogen derived from renewable resources.*

The phrase 'new renewable energy' as opposed to the more general 'renewable energy' has started to emerge in some policy circles. This has occurred for several reasons. First, it identifies the latest wave of renewable technologies, such as leading-edge silicon technology and advances in fluid mechanics and composites. Second, it distinguishes the set of renewable technologies and sources that are environmentally safe and sustainable.

In practice this is intended to isolate technologies such as large hydroelectric dams, which have major local environmental impacts and which can also result in significant greenhouse gas emissions.[3] The vexed nature of the dams debate is captured by the 2000 report by the World Commission on Dams (WCD, 2000):

> *While the immediate benefits were widely believed sufficient to justify the enormous investments made – total investment in large dams worldwide is estimated at more than $2 trillion – secondary and tertiary benefits were also often cited. These included food security considerations, local employment and skills development, rural electrification and the expansion of physical and social infrastructure such as roads and schools. The benefits were regarded as self-evident, while the construction and operational costs tended to be limited to economic and financial considerations that justified dams as a highly competitive option.*
>
> *As experience accumulated and better information on the performance and consequences of dams became available, the full cost of large dams began to emerge as a serious public concern. Driven by information on the impacts of dams on people, river basins and ecosystems, as well as their economic performance, opposition began to grow. Debate and controversy initially focused on specific dams and their local impacts. Gradually these locally driven conflicts evolved into a global debate about the costs and benefits of dams. Global estimates of the magnitude of impacts include some 40–80 million people displaced by dams while 60 per cent of the world's rivers have been affected by dams and diversions. The nature and magnitude of the impacts of dams on affected communities and on the environment have now become established as key issues in the debate.*

However, assuming any technology to be without side effects is simplistic and needs to be treated with caution lest we establish expectations that are too high. Further debate regarding large hydroelectric development is practically beyond the scope of this book, since it is not a technology that needs to be leveraged into the market any more but is a well-established 'conventional' power source.

While no formal definition would suit all commentators, a modern list of new renewable technologies might include wind power, solar photovoltaics, solar thermal electric, small hydroelectric plant, sustainable biomass, wave, tidal and ocean current systems, and geothermal sources.

Do the policies avoid putting industries which are extremely different in size and maturity into direct competition?

Pitching any small embryonic technology or industry against a large established industry is unlikely to succeed. Unless the David has some remarkable attributes, the Goliath usually wins. (A quick look at computer software tells us that market power is a mighty beast.) A revolution in renewable energy pricing is underway today, but it would not have happened nor will it continue unless renewables are provided with specifically defined support and permitted to access markets with which to increase production and thus drive down the costs.

This book's goal is not to win arguments about why renewables are better than nuclear power, coal plants, geosequestration or large hydroelectric stations. Rather it is intended as an examination of renewable energy specifically. However, allowing larger industries access to the development mechanisms of the renewable energy industry is fraught with risk to the new renewable providers. Although diversity may be a choice policy-makers intend, there are clearly unseen consequences inherent in large, market-dominant technologies accessing resources which are set aside for developing industries.

For example, a policy-maker may indeed wish to support nuclear power, large dams or geosequestration. However, the question is, are adapted or technology-neutral renewable energy policies the best way to do this? For example, a single nuclear power station uses up the equivalent energy market share of 30 or more renewable power plants. This uses up both electrical load and financial support.

As well as nuclear power, there is significant pressure in some circles for coal plant carbon capture and storage to be considered zero emission or even 'renewable'. As another example, there are (and indeed it would be impossible to proceed if there were not) various other funds available to get a demonstration geosequestration plant up and running and, as such, the plant would also not be competing on a level basis with renewables. While the policy focus may well be on the sequestration side, the coal plant would be of a size that could monopolize resources capable of providing several years of renewable energy development.

Myriad inequalities exist between large established industries and embryonic ones. These must be considered when expecting them to compete and indeed when comparing their performance. Experience indicates that the best outcomes are always achieved when large established industries are not mixed in the same schemes as new renewable energy industries.

Are the policies focused on renewable energy free of contradictory goals?

We have established that there is a need for very clear objectives, and so it is important these are not lost in the interests of simplified policy-making when it comes to defining eligible technologies or resources.

The question 'Why have a renewable energy policy at all?' is often raised by politicians, with the proposal that there are other more market-friendly alternatives to achieve the same goals. For example, a market scheme for climate-friendly technologies might have a mix of renewables, gas plants, nuclear power, energy efficiency and carbon sequestration with tree planting. Such a so-called flexible scheme clearly comes under the heading of carbon trading and assumes some carbon taxation or capping.

However, as we have noted before, many energy efficiency measures are cost effective in their own right and so would trade at a low or zero price with regard to carbon trading. Other schemes such as tree planting are cheap (and sometimes have questionable effectiveness – for example, who is liable for the carbon that is no longer sequestered if the plantation's trees burn in a forest fire?). Under such conditions, renewable energy plants would not be competitive and so would not be built, and therefore clearly this would not be an effective renewable energy development scheme.

In practice it is quite possible to have clean energy legislation that creates various markets for carbon and sees money change hands, but does little or nothing to change business as usual on the generation side. Therefore such legislation does nothing to prevent release of mineral carbon into the atmosphere and biosphere. In the long term, the 'low-hanging fruit' of tree planting and other low-cost, non-energy-related greenhouse mitigation will be used up and the more costly renewables will become competitive in due course. In the meantime, however, renewables are effectively put on hold and any intended development objectives of the renewable energy industry are not delivered.

Therefore, the more specific a policy is about the technologies eligible for support under a scheme, the more certain will be the delivery of those technologies into the market, and the more rapid the delivery of the associated benefits/objectives. The crucial element is to ensure that renewables sit at the centre of the policy if they are indeed the goal.

Are technologies treated differently if necessary? And how?

There is probably no such thing as technology neutrality because technology choices are inherent in strategic decision-making about energy. Nevertheless a potentially appropriate level of neutrality may be created by identifying technology groups with defining characteristics that render them suitable for specific policy targets.

Renewable energy can be assigned to vastly different groups. Yet the more specific the definition of the group, the more targeted can be the policies. Energy can be defined from the point of view of emissions, for example high emissions (such as coal), low emissions (gas) and zero emissions (renewables).

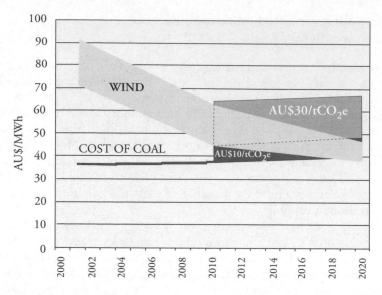

Source: International Energy Agency

Figure 3.1 *The convergence process: Wind undercuts coal prices in Australia with carbon prices at AU$30 and then AU$10/tonne of CO_2 emitted*

Technologies can also be defined from the point of view of economics (with or without factoring in the price of carbon). For example, we can classify energy technologies as low-cost commercial, high-cost commercial, non-commercial but with declining prices, and non-commercial without declining prices. Obviously, energy technologies can also be defined by resource, for example, fossil fuel, nuclear, solar based, wind based, biomass based, or sea/ocean based. Finally, they can be defined by the technology itself, such as wind onshore, wind offshore, PV solar thermal electric, and solar thermal hot water, to name just a few.

Policy-makers reading this book may not be in a position to make decisions about the introduction of carbon taxes or emissions caps. However, they may be aware that in 10 or 20 years international carbon constraints look likely. Therefore they can make strategic decisions about the relative importance of renewable energy portfolios and other technology groups, and how the relative costs are likely to evolve.

Australia has some of the cheapest energy in the world, yet in Figure 3.1 we see how renewables such as wind, which are commercially maturing but still expensive today, would become cheaper than coal generation under carbon caps. We also see that coal has upward price pressures whereas wind will still experience the significant price decline linked to industry growth.

Thus a decision-maker may see merit in building up some industry groups or technologies, putting others into R&D or commercialization programmes and choosing to leave alone others which have less long-term viability.

Within a given group even more specificity may be desired, for example along the lines of the type of resources being harnessed or even the type of harnessing.

Table 3.1 *The NFFO technology mix (see Chapter 6 on the UK)*

Technology	Contracted Projects
Landfill gas	329
Wind	302
Hydro	146
Municipal industrial waste	90
Biomass	32
Sewage gas	31
Wave	3

Source: UK Department of Trade and Industry

The technology breakdown used in the Non-Fossil Fuel Obligation (NFFO) (see Table 3.1) was one of the first examples of this degree of specificity. Its intention was to maximize the industry development of each area, rather than having one technology stall as another took the lion's share of resources.

Is the intended technology or mix of technologies properly articulated?

With policies geared specifically towards renewable project implementation, a spectrum of desired outcomes may be possible. At one end of the spectrum there may be a single technology policy, for example, to support solar hot water heating, while at the other end of the spectrum there may be a mix of possible technologies. In any case, the range of technologies to which the policy applies must be clearly defined in some way.

The advantage of being technology specific is that it becomes possible to get what is wanted in terms of technology and industry development, and can focus resources accordingly. However, the general criticism of such specificity is that it is not as financially optimal as a policy under which technologies and projects compete on equal footing.

The portfolio approach provides for a set of pre-defined technologies to compete for resources or subdivided market share, either on an equal basis or with some level of weighting applied to provide more targeted incentives.

The most common technology-neutral approach is to avoid specifying the technology at all, and instead to define eligibility criteria. In this model one might expect criteria such as 'zero-emission standards' or 'non-fossil fuels'. In principle these have an appeal of elegance and simplicity and thus let every flower bloom or allow the fittest to survive.

We have discussed the flaws in a wide-open technology-neutral approach, but limited neutral definitions are also open to interpretation. For example, the UK Non-Fossil Fuel Obligation was actually set up to support nuclear power rather

Table 3.2 *The range of outcomes possible with single, mixed and neutral approaches to technology selection*

	Single technology	Technology mix	Technology neutral
Sustainability objectives	YES	YES	YES
Energy policy reform	Depends on the technology	YES	YES
Greenhouse gas mitigation	Depends on the technology	YES	YES
Energy-cost and least-cost planning (internalization)	NO	NO	YES
Energy security	NO	YES	YES
New industry/manufacturing	YES	YES	NO
New intellectual property	YES	YES	NO
Job creation	YES	YES	NO
Rural investment	Depends on the technology	YES	YES
Nuclear phase-out	Depends on the technology	YES	YES

than renewable energy. Increasingly we are seeing pressure for carbon geosequestration to be considered renewable or zero emission. So we must note here that definitions of neutrality must be made with full knowledge of the technologies they will actually end up supporting.

Even if we identify renewables as a specific group, we must still choose between the three approaches of single, portfolio or neutral. For example, identifying a specific technology and providing a targeted push is likely to pay dividends in industry development, job creation and exports as it has done in Denmark with regard to wind power, Japan with PV and Israel with solar hot water heating. However, it will not provide a broad base of energy production, nor will it rapidly provide the most energy for the least cost. Alternatively a technology-neutral approach is likely to provide the greatest renewable energy production for the least cost, but it may not encourage manufacturing or diversity. In between is a mixed approach, the performance of which will depend on the nature and extent of the mix and measures (see Table 3.2).

Are the consequences of including any environmentally unsustainable technologies fully understood?

Finally, experience indicates that the inclusion of environmentally unsustainable technology or resources under a renewable energy policy can lead to large social and therefore political problems.

The most prominent areas of concern are new, large hydroelectric projects and unsustainable biomass. The former is risky because of impacts caused by flooding and river-flow disruption, the latter if biomass acquisition leads to environmental damage and/or a loss of biodiversity.

There is a risk that a policy framework may fall victim to a civic backlash if it has sustainability as one of its key motivations but is seen to be exploited for outcomes that are non-sustainable or result in a net loss for the environment.

Appropriately Applied Incentives

Let us assume that we have now fully defined the resources and/or technologies eligible under a policy. The policy framework as a whole must then be fine-tuned to maximize the long-term harnessing of the spectrum of the country's indigenous renewable resources. The following questions and answers attempt to cover the challenges we may meet on this front.

Is there flexibility to allow new technologies to be included and to evolve in future?

Policies intended to support renewables must permit technologies to be introduced, to migrate as they evolve and to ultimately depart, leaving support focused where needed.

Renewable energy sources are largely known today. The technologies to harness them are relatively new compared to thermal energy production – which is two centuries old. These renewable technologies are still evolving rapidly and occasionally make major leaps.

As currently non-commercial technologies become more viable, they will need to migrate from R&D or demonstration support to the commercial technology bracket. In the latter bracket the focus is on industry development using some type of policy to make up the price gap. As the price gap finally narrows, these technologies may be capable of migrating into the broad energy market which has carbon constraints. In some cases, renewables are also cost effective in a non-carbon constrained market as is, but these cases tend to niches.

Therefore, policy needs to allow new technologies to join the list of new renewables being developed, and thereafter to migrate between support schemes as their costs decline and volume increases. The Spanish case study in this book illustrates how failure to build in this policy feature initially blocked solar thermal electricity's migration from demonstration to industry development, holding back a country-appropriate technology.

Are the identified technologies actually being developed by the policies?

Once we have identified the technologies or groups that are being targeted we need to ensure that these technologies are indeed being developed.

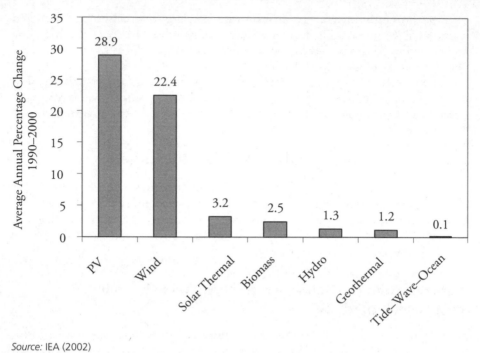

Source: IEA (2002)

Figure 3.2: *Growth rates over the decade 1990–2000 show how PV and wind energy surged while almost all other renewables languished*

Deep GHG emissions cuts require a portfolio of zero- or low-emission sources to replace the current carbon-intensive energy supply. How long that process takes depends in part upon on the time required to install and commission industries.

A time-critical element is therefore the industry growth rate. For example, wind power has grown at more than 25 per cent per annum over a decade and solar PV has had similarly spectacular growth. But are the other renewable industries performing as well?

Here are the four main reasons technologies can be left behind:

1 First, drivers may not be optimized. The drivers may need to be quite different for different technologies, varying substantially for wind, small hydro, bio-fuels and solar PV.
2 There may be variations in financial efficiency. The technologies will require different levels of support due to their different levels of maturity. If this need is ignored, the market may come to be dominated by the cheapest technology alone.
3 Parallel industry development may not occur. Different technologies will have a varying capacity to deliver, based on different industry sizes and pricing. A one-size-suits-all policy may render many technologies dormant until prices become suitable, and this wastes valuable development time.

4   There may be inadequate or unclear policy signals. Clearly identifying the technologies indicates to decision-makers of other ministerial or departmental portfolios (for example, treasury, agriculture, environment or planning) which other areas of policy must also be modified. For example, grid connection policies for wind or household metering for PV may need revision.

The key point is that if deep GHG reductions or long-term industry developments are important and that therefore the aim is to deliver a suite of industries, the policy framework must ensure that the country-appropriate technologies it identifies are not left to languish or stall.

Is the line drawn around the eligible technologies in a way that allows adequate promotion for the country-appropriate technologies?

Countries that promote the most globally successful renewable technologies may also fail to incubate technologies that will harness abundant national renewable resources that are currently untapped.

Why are Polynesians excellent sailors, Russians good horsemen and Arabians' architecture so rich in passive solar techniques? Wind power has become a huge success, but almost entirely through perseverance and technological development in countries like Denmark, Germany and the UK. The fact that modern wind technology emerged from these countries is no accident. They are windy places.

Vast areas of the planet are bathed in strong sunlight. While solar thermal technologies may have fallen behind in cooler northern countries, they should not be ignored by nations assessing their own indigenous resources. The take-home message is that the appropriate mix of technologies to be implemented or developed will always vary from country to country. And if a technology has not reached adequate maturity, there exists an opportunity for a future national industry and even internationally valuable intellectual property.

Adequacy

New renewable markets require a strong injection of resources to get them running, after which they can be optimized for smooth performance. There are many types of drivers. The adequacy test must be applied to ensure that whatever its type, the driver is providing enough fuel for renewable development. To reiterate, the driver is what starts the engine, and it must initially be a fuel-rich mix which can be made leaner in the longer term.

Since manufacturing and job creation is often the jewel in the crown of a renewable energy policy, it is very important to achieve the correct financing, duration and intensity thresholds. These sometimes run counter to short-term economic efficiency which can create arguments for minimizing the cost impact of renewables or waiting until the technology undergoes further price reductions.

Table 3.3 *Estimate of the future cost reduction of wind-generated electricity*

Source	Relative share (%)
Design improvements – weight ratios	35
Improved conversion efficiency[4] – aerodynamic and electric efficiency	5
Economy of scale – steady serial production and optimization of logistics	50
Other contributions – foundations, grid connection, O&M [5]	10
Total	100

Note: Percentages refer to a total reduction of 15 per cent from 2000 to 2004. O&M: operation and maintenance.
Source: Mallon and Reardon (2004)

However, Table 3.3 shows how more than 50 per cent of cost reductions are achieved through economies of scale production and learning by doing – illustrating that the rationale to sit and wait holds little water. The empirical evidence clearly shows that firm and decisive policy brings manufacturing, jobs and exports. This begs the question of whether it is better to spend less money, only to have more money leave the country, or to spend more money up front in order to achieve greater domestic retention and manufacturing activity. I leave it to the economists to argue this point.

When we examine adequacy, there are at least three important questions to answer, and these will now be discussed in the following subsections.

Are the policies leveraging private sector investment?

Simply throwing money at renewables does not work[6] – governments do not have deep enough pockets. The way to mobilize renewables is to mobilize private sector finance. And the key to that is ensuring investors can get a return on their money equivalent to or better than that available elsewhere. Good policy design will need to consider the investment profile of renewable projects and determine what is required to make the industry attractive to private investment.

An excellent example is the early investment by the US government in wind energy in the wake of the 1970s oil shocks. The result was a series of massive experimental wind turbines produced by companies like Boeing and Westinghouse in the mid-1980s. However, it was primarily Danish and German companies (including Vestas, an agricultural machinery manufacturer) that eventually prevailed, building up a business in a stable policy environment that allowed the steady development of a wind industry.

Are the returns on investment comparable with other alternatives?

Renewable energy forms are cost competitive with conventional sources in an increasing number of jurisdictions. Intervention may not be required in such cases but, for bulk energy in large economies, some level of pollution internalization or positive intervention is necessary. We can think of positive intervention as compensatory, recognizing that it is easier to compensate 10 per cent of the energy sector for market distortions than to charge 90 per cent of the sector for pollution.

By way of example, I recently undertook a renewable energy scoping exercise on the Pacific island nation of Niue. There the cost of electricity to consumers is very high at about 30 New Zealand (NZ) cents (approximately 22 US cents) per kWh, a price which includes a 50 per cent state subsidy! By comparison, estimates show wind energy on the island could be produced at 20 NZ (15 US) cents. In Niue's case, the economics of electricity prices will clearly not be the limiting factor for the implementation of wind power. At the opposite end of the spectrum lie countries such as Australia where electricity is generated at a mere 1.5 US cents per kWh. There, policy intervention is needed to make up the renewable energy cost gap.

Interventions which create some sort of direct or indirect price support must be implemented at levels that allow *independent*, commercially viable projects to proceed. At first glance, this level would appear to be the difference between the price of conventional generation in the national systems and the price of new renewables. However, the price of either may not be what they first appear. The word 'independent' is very important here. It refers to projects which can be externally financed through normal corporate lending based on their own merits (project financed) – at financing costs which may be significantly higher than those which state-owned entities or large corporations may be able to achieve.

To explain this point, we need to understand that many large or state-owned companies are able to finance projects internally or off a balance sheet. That is, they can guarantee to make the repayments based on company cash flows as a whole, not the project itself. This may allow for better financial rates which permit projects to become viable. For example, a satisfactory internal rate of return for a state-owned company might be 6 per cent. However, institutional lenders might demand a return of 12 per cent (depending on the value placed on a green project), while for private financiers the rate might be up to 17 per cent, and for venture capital companies it could be 35 per cent. So the smaller renewable energy players must bank the project on the merits of the project alone and pay substantially more for their equity.

A support level which enables only state companies and large corporations to develop renewable energy is false economy. It excludes the wider private development sector and thus reduces the incubation of a highly competitive private development industry.

Are the investment periods long enough for project investors to get a return on equity?

Equity participants are very sensitive to the duration of projects. Almost all renewable energy projects are capital intensive and have low fuel costs. The projects are often financed through a mixture of equity and debt. Over the project's operational lifetime, its revenues are often first used to pay off the debt and thereafter the owners obtain a return on their equity. Projects with a lifetime of 20 years require a very long-term commitment by investors.

Table 3.4 *Industry development rates and manufacturing as a function of the size of the Mandatory Renewable Energy Target (MRET) in Australia*

	Current MRET	5 per cent MRET by 2010	10 per cent MRET by 2010
Total renewable energy target (GWh)	9500	19,000	30,150
Total wind in target (GWh)	2470	7220	12,795
Total installed wind capacity at 2010 (MW)	940	2842	5037
Annual turbine installations to 2010 (based on an average wind turbine size of 1.5MW)	90	270	480
Blade manufacturing facilities at threshold demand (based on a threshold of 100 wind turbines per year per blade manufacturer)	<1	2–3	4–5
Maximum number of facilities (accounting for market share)	<1	2	4

Source: Maddox (2004); produced for the Australian Wind Energy Association

If a project is to be viable, a balance must therefore be struck between the duration of a project and the cost of energy. Ideally a project would receive support for its full lifetime. If the support falls below this ideal, the project must charge higher prices for the energy in order to recover the debt and equity.

The use of excessively short periods of support can lead to prices which do not represent the real cost of the renewable energy and therefore reduce its competitiveness. This means that unnecessary costs are passed onto consumers or the taxpayer.

Are the installation intensity and scheme duration adequate to enable manufacturing?

The manufacturing life cycle is much more long-term than that of installed projects. It depends on a sustained stream of projects purchasing hardware at a sufficient rate over long enough periods for a factory to pay for itself and deliver decent profits. If this sustained installation period is absent or if the throughput is too small, then it may be cheaper and less risky for suppliers to import technology and domestic manufacturing will not occur.

Table 3.4 shows how different industry development rates are used to determine the number of manufacturing plants likely to be established under a given scheme. It is important to remember that the market volume must be divided between the likely number of manufacturers, and in a competitive market this will be at least two and preferably three. Therefore the optimum market size needs to be three times the volume required for a single manufacturing plant.

Note: Kenetech machines dominated the global wind market in the early 1990s, but failed to cope with the severe US and then Indian boom–bust cycles.
Source: Paul Gipe

Figure 3.3 *Kenetech wind turbines*

Stability

The final, but most important driver-based measure is the need for stability. As seen in the previous chapter, an unstable policy is worse than no policy at all. Why? Because an unstable policy leads business to make investments that may collapse as future support wanes. A sector of formerly enthusiastic industry participants may be deterred from engaging in a further round. They may prefer not to engage until the government indicates it is ready for a sustained commitment.

Policy stability is a fundamental requirement of market certainty, which in turn is strongly related to both the production price of renewable energy and the development of manufacturing capacity.

Stability may be considered the key ingredient in the success of the German feed-in law, which remained largely unchanged for 10 years. Conversely, Swisher and Porter describe in Chapter 7 on the US how policy instability characterized the early US approach to renewables and continues to define the ongoing football game developers there must play with production tax credits. The Swisher and Porter comparison of US and German policies later in the book provides valuable insights.

The stability of support in Germany allowed project developers to be sure of prices and also to develop projects in a steady stream. Furthermore, the increasing experience and concomitant industry sophistication – in projects, financing and technology – has led to lower costs. These are benefits that can allow future revisions to bring down costs and ensure economic efficiency.

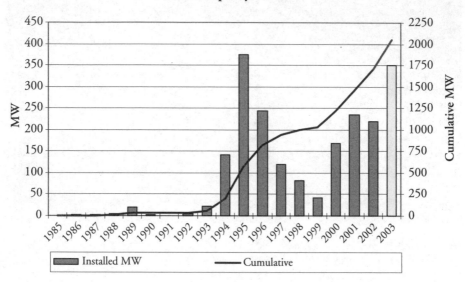

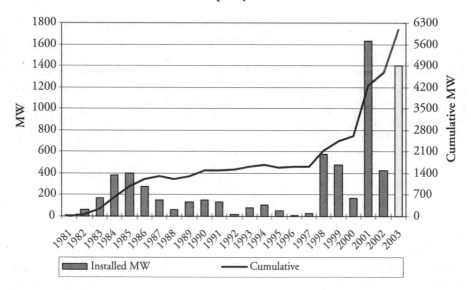

Note: Data for 2003 are forecast
Source: BTM (2004)

Figure 3.4 *Illustration of the boom–bust cycle in India and the US*

So what does stability mean in practice? The following discussion attempts to cover some of the most important considerations.

Does the policy framework avoid boom–bust cycles?

The boom–bust phenomenon constitutes what is probably one of the most common failures in renewable energy frameworks.

Only a very small number of countries have successfully avoided this pitfall. They have done so with frameworks that avoid excessive time pressures or excessive competition for funds. We could describe their policy frameworks as having a more 'open' architecture as they tend to attract rather than deter.

Many treasury economists are worried by targets and spending volumes or periods which are not fully pre-defined, and so caps are usually put in place. These caps are often couched as aspirational targets such as 'one hundred thousand roofs', 'a thousand megawatts', or 'a billion litres'. This approach has its merits, but what happens in successful cases, when the target is reached or the money spent? The target of course becomes a cap and the boom becomes a bust.

As I write this chapter, I have just returned from several days of lobbying the Australian parliament about the current status of wind power. I alerted them to the problem that wind in Australia is a victim of its own success. The Australian federal MRET policy has, in principle, a duration of 20 years. However, all of the project installations required to meet the target will have been completed within the first five years of the policy period. Projects have been found, factories built, people employed, farmers' land leased and so on. Yet once those five years have passed, the industry is poised to fall off the investment cliff into a market and policy void.

Boom–bust cycles can result from several factors, four of which I will describe here:

1 There may be a time restriction in which schemes are limited in terms of time or eligibility period. This leads to rushes of development and then stalls, as has been evidenced under the system of US tax credits.
2 There may be limited pools of resources that are unpredictable in amount or spent on an irregular basis. This leads to floods of development when there is incentive and droughts when there is not, typified by the pattern witnessed under the UK NFFO policy.
3 Targets or caps may result in development rushes to meet the said targets followed by a rather abrupt halt, as we have discussed is underway in Australia.
4 Excessive competition may create an environment in which developers must race one another to be included in a scheme.

Does the policy plan for the whole cycle of an industry's development?

Long-term and stable policies must entail some sense of where these industries are headed. Such policies must also profile the support that will be required for the industry's build-up and complete the process with an exit/integration strategy.

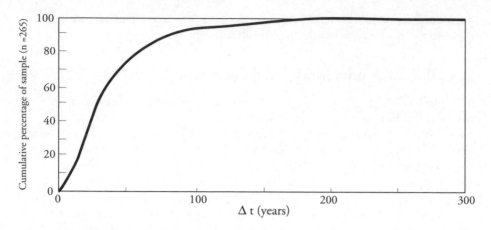

Note: The mean technology diffusion rate for the sample was 41 years.
Source: Grubler et al (1999)

Figure 3.5: *Technology diffusion rates for over 265 different technologies all following the same path*

It is no minor task to manage sustained and steady development for industries which possess spectacular growth rates. Getting it right results in a success like Nokia; getting it wrong results in the renewable equivalent of a dot-com bubble. Thankfully renewables are not the first to tread this path. We can therefore identify the standard growth path expected for these industries and plan their trajectories.

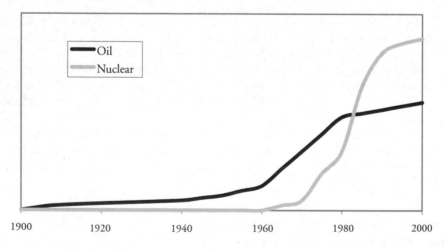

Sources: Nuclear capacity: McDonald (2004);
Oil barrels produced per year: Meling

Figure 3.6 *Growth curves of two energy industries (non-dimensional)*

In so doing, we can identify three fairly distinct areas of the growth phase. The slow, flat start is witnessed as new industries materialize from nothing and attempt to make sense of the world. Then we see the J-curve, as the industries get progressively better and bigger, leading to exponential growth. Finally we see a plateau as resource or other constraints gradually confront the industry.

Often we see policy measures that keep industries limping along in the first part of that curve (the engine that will not start). Or we may get the industry onto the exponential growth phase of the curve but without an idea of how long growth is to be sustained or financed, leading to the painful booms and busts.

Understanding the expected future role of a technology or resource base permits scenarios of the long-term trajectory to be established. With this understood, the plan to actively manage that trajectory and the resources required can be deduced and integrated into the policy framework

Does the policy provide an ongoing steady pull on development?

The opposite of a boom–bust cycle might best be referred to as a 'steady pull', with growth following most closely the S-curve of industry development. However, a 'steady pull' in this instance in fact produces dynamic growth. At some stages, linear growth may be considered steady, at others growth should be exponential and, as the industry reaches full maturity, a gradual levelling off is appropriate.

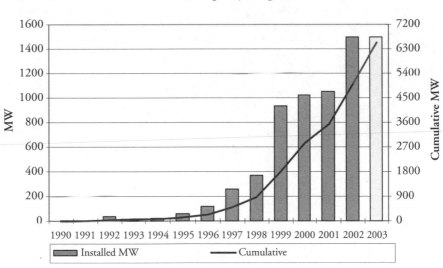

Installed capacity in Spain

Note: Data for 2003 are forecast
Source: BTM

Figure 3.7 *7000MW of wind capacity were installed in Spain over 10 years, without any market contractions*

This is why the best measure of steady pull is the S-curve as a whole. The fewer wobbles and deviations from the curve, the more steady the pull.

Thankfully, this path is not as hard to achieve as it may at first appear. If the price behaviour of renewables did not change, it might be impossible to manage or coordinate such dynamic market behaviour, but in fact it is the price decline of any product that can itself result in this S-curve. Thus, the natural industry growth will be a steady pull along the S-curve, provided the policy framework does not artificially introduce perturbations (and barring unexpected disruptions from other sources).

The example in Box 3.3 shows how the exponential behaviour spontaneously emerges provided the resources made available for the incentive are maintained at the same level. This phenomenon occurs not because of the price decline per se but rather because of price convergence. In theory, if price convergence is achieved or is surpassed, the new technology will take up 100 per cent of market share. In practice, other constraints come to prevail and the full costs of renewable energy will not continue to decline forever. Market-size external obstacles and the costs of applying solutions ultimately curb the growth.

Box 3.3 *The S-curve by numbers*

Initially, as the market for a new technology expands, money spent does not go very far as the technology is expensive. However, the more that is produced, the cheaper the product becomes, allowing the same amount of money to go much farther. Figure 3.8 shows the price of a renewable technology declining with time. The price support goes towards making up the cost difference between normal technology costs and the new renewable technology. As the figure shows, this cost difference gradually declines towards zero.

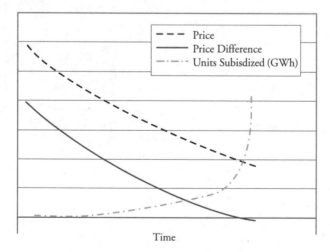

Figure 3.8 *The decline of price and price difference, and the resulting effect of a fixed level of financial support on volume of commodity*

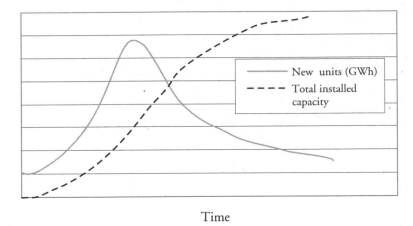

Time

Figure 3.9 *Typical curves comparing the number of new units sold and total installed capacity*

The above assumes that price convergence can be achieved for one or all renewable energy technologies. If it is achieved for a given technology, the incentives ultimately become unnecessary. On the other hand, there may be some technologies that reach a price plateau preventing them from directly competing with fossil fuels. In this case, the price of carbon or other externalities play a role in determining their ongoing viability.

As the cost declines, the number of units that can be subsidized by a given volume of support rises exponentially. A new technology dominates the market not when its price is extremely low, but rather when its price is simply lower that its competitor. It is similar to the tale of two explorers in the African jungle who find themselves stalked by a leopard. As one dons running shoes his fellow says, 'That's a waste of time; you can't run faster than a leopard.' The first replies, 'I don't have to run faster than the leopard. I only have to run faster than you.'

There will be limits, however: limits on resource availability, for example, fuel or site cost increases, secondary costs such as managing peak load, and the cost of storage and distribution. As these constraints are felt, the level of new installation tails off and the activity required to maintain and replace that installed capacity sustains the industry thereafter (see Figure 3.9).

Is the resource base for the incentive sustainable?

In the real world no financial incentive is open-ended because no resource is inexhaustible. The additional cost of renewables is usually passed onto the consumer but sometimes it comes out of the tax pool. However, time after time, schemes intended to be open-ended are suddenly halted by intervention if they are running too fast for the supporting resource base to cope.

Working in favour of renewable industry expansion is the typical coupling of steady growth with steady price declines – the so-called learning curve. Steadi

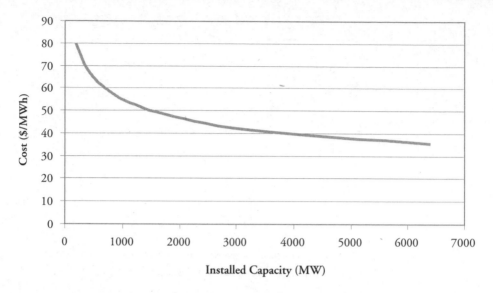

Figure 3.10 *Standard graph of price decline versus capacity increase for a learning rate of 15 per cent per doubling in installed capacity*

ness is important here since short-lived spikes tend to push up prices because they result in high demand combined with supply shortages.

So whatever the policy driver used to provide the incentive, the aim must be to achieve the right balance to provide a steady draw on renewable industry development. This steady pull should be matched by a balanced and sustainable pull on the resources made available by the government and by the wider economy.

One important factor that must also be considered concerns the resource base. A renewable industry contributes to the economy itself – investment is made, factories are built, people are employed and taxes are paid. So the flow of investment is not simply one way, but rather two way. This reality must be reflected in the equation of resource demand from the wider economy. Quantifying this effect is too wide a topic to consider here but it can be performed by independent economists (see Figure 3.11).

Are there long-term energy policies that provide guidance and surety for evolving policies and measures?

Because policy frameworks evolve, we must ensure that the revision of renewable and general energy-sector policies will not destabilize the industry.

In the words of the wise, 'All things change and nothing is permanent.' Therefore the policy framework must have some faculty to handle change. The more we try to wrap a policy in iron to shield it from change, the more we limit its ability to evolve to a changing world. We must reasonably expect that the policies we make today will be changed by others in the future.

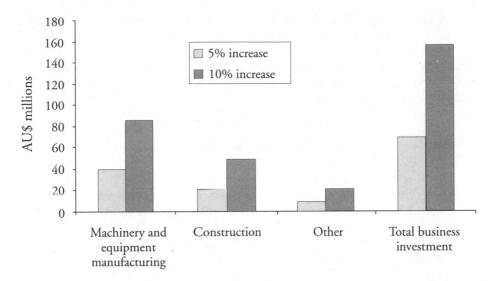

Source: CRA (2003)

Figure 3.11 *Modelling of change in business investment from increases to the renewable energy targets in Australia*

This said, people often resist change just as they fear the unknown, hence the old maxim: 'If it isn't broken, don't fix it.' Although a time may come when a particular policy requires fixing, we must hope that reform reflects the original policy objectives. This is why it is wise to reflect the objectives themselves within the policy.

The types of policies most applicable for the long term are non-legislated statements of objective. The most pertinent current example is the UK government policy for 60 per cent CO_2 emissions reductions by 2050:

> *Our ambition is for the world's developed economies to cut emissions of greenhouse gases by 60 per cent by around 2050. We therefore accept the Royal Commission on Environmental Pollution's (RCEP's) recommendation that the UK should put itself on a path towards a reduction in carbon dioxide emissions of some 60 per cent from current levels by about 2050. Until now the UK's energy policy has not paid enough attention to environmental problems. Our new energy policy will ensure that energy, the environment and economic growth are properly and sustainably integrated. In this White Paper, we set out the first steps to achieving this goal.* (DTI, 2003)

Although this goal cannot be set as a single piece of legislation in any meaningful way, it does demand that whatever legislation is enacted be consistent with this objective. Unless climate change is solved by some great, yet-to-be-discovered innovation, the objective is likely to be just as pertinent 20 years from today – perhaps more so.

Consequently as the UK reviews and evolves its renewable energy policies they will all (one hopes) be in keeping with this overarching climate policy objective.

More robust still for renewables would be to have the stated objectives of long-term renewable targets enshrined in government policy. Given the slow turnover time of energy production facilities, such targets will preferably extend for several decades or even half a century.

Contextual Frameworks

It takes more than a driver to deliver a healthy renewable energy market. Depending on the technology or energy source in question, a range of policy instruments or adjustments are required to integrate renewable projects and the energy they produce into existing systems. In any given country, it is important to have an overview of where the current and potential obstacles are, and therefore the interventions required. The questions and answers considered in the following sections should provide some guidance on this count.

Is there an overarching national policy objective in place to guide all policy-making?

A single national objective for specific long-term renewable energy growth or emission reductions provides three important foundations: an incentive for positive legislation; a reason to review existing legislation that may be acting as a barrier; and insurance against obstructive legislation being put in place thereafter.

In practice it can prove to be very difficult to establish cohesive policy frameworks unless such objectives are stated and committed to by governments.

Once an overarching objective such as this is in place, there are two ways to ascertain that the correct policy frameworks are present. First, we can take a slice through the policy framework from a project perspective. And second, we can drill down through all the levels of government and their legislative jurisdiction. The following two questions are designed to illuminate these two paths.

Have all the laws that affect renewable energy projects been checked to ensure they are conducive and not obstructive?

In order to identify the supporting frameworks that must be established, we should again put ourselves in the shoes of the business person attempting to establish renewable production and travel the path from project conception to project completion. We need to check every point that the project hits a contractual or legislative interaction to ensure the policies are conducive and not obstructive to renewable energy development (see Figure 3.12).

	Technical & Commercial	Environmental	Dialogue & Consultation
Stage 1 SITE IDENTIFICATION	• Preliminary Resource Assessment • Grid Availability • Access • Site Ownership Pre-Feasibility Study	• Land Use (Macro) / Classification • Fatal Flaw Review	• Local / State Planning Authority • Network Service Providers
Stage 2 SITE ACQUISITION & ASSESSMENT	• Site Acquisition • On-Site Wind Monitoring • Resource Modelling	• Visual Effects • Proximity to Dwellings • Land Use (Site Specific)	• Landowner Consultation • Wind Rights • Local Community • Planning Authorities
Stage 3 PROJECT FEASIBILITY	• Design & Estimated Energy Yield • Energy / Green Tariffs • Electrical Connection • Risk Assessment • Project Costs • Plant Supplier(s)	• Environmental Approval Requirement • Scoping Document	• Power Purchasers • Network Service Providers • Planning Authorities • Local Community
Stage 4 DA & EIA PREPARATION	• Wind Farm Layout • Transmission Line Route • Turbine Noise	• As per legislative requirements [Local, State and Federal]	• Landowner Consultation • Local Planning Authority • Community Groups • Local Community
MILESTONE: DEVELOPMENT APPROVAL			
Stage 5 DESIGN & TECHNICAL DEVELOPMENT	• Wind Resource & Energy Yield Refinement • Wind Farm Layout • Transmission & Grid Connection	• In accordance with the terms and condition as specified within the DA approval documentation.	• Landowner Consultation • Local Community • Community Groups • Network Service Providers
Stage 6 COMMERCIAL ENGINEERING	• Project Finance • PPA Finalization • Turbine Selection • Contract Packaging • Due Diligence • Owner's Engineer • Insurance	• None	• Banks / Financial Institutions • Power Purchasers • Network Service Providers
MILESTONE: FINANCIAL CLOSURE			
Stage 7 CONSTRUCTION	• Site Management • Construction Standards • Operational health & safety	• Environmental management, supervision and auditing in accordance with the terms and conditions as specified within the DA approval documentation and Environment Management Plan.	• Landowner Consultation • Local Planning Authority • Local Community • General Public • Community Groups • Network Service Providers
Stage 8 PROJECT COMMISSIONING	• In accordance with pre-agreed commissioning and project acceptance procedures.	• None	• Site Management • Local Community • Community Groups • Network Service Providers
MILESTONE: START OF COMMERCIAL OPERATION			
Stage 9 OPERATION & MAINTENANCE	• In accordance with pre-agreed operation and maintenance procedures.	• Environmental Monitoring • Complaint Procedures	• Landowner Consultation • Local Community • General Public
Stage 10 DECOMMISSIONING & REINSTATEMENT	• In accordance with the terms and conditions as specified within the DA approval documentation and Environment Management Plan.	• Land Reinstatement	• Landowner Consultation • Local Planning Authority • Removal of Legal Obligations

Note: DA = development application; PPP = power purchase agreement
Source: Based on British Wind Energy Association sources and information provided by commercial project developers

Figure 3.12 *Diagram of the project cycle*

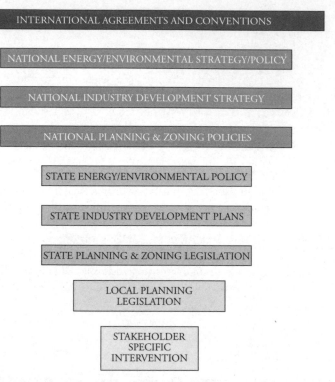

Figure 3.13 *Tiers of policies, plans and legislation that may be used to deliver or facilitate renewable projects*

Are the policies and measures at all tiers of government consistent and self-reinforcing?

Government policies with respect to large-scale renewables can come at a variety of levels. Figure 3.13 illustrates a sample cascade of legislative tiers.

Beginning from the highest level, legislation may start with international agreements, such as the EU Renewable Energy Directive or the Kyoto Protocol. Then there are national energy policy-making and state, provincial or other categories of regionally based initiatives. Finally comes local government decision-making.

Each area must be examined at the outset to assess its renewable friendliness. As with market drivers, the more levels of cooperation and dissemination that exist, the more effective the policies are likely to be. It is crucial that the policy framework as a whole be free of areas of policy setting which are mutually inconsistent or guided by contrary principles.

Energy Market Reform

Some renewables will leapfrog past conventional generation in locations which lack access to commercial energy. For the most part, however, renewables will be integrated into physical systems and management systems which have not been designed for this type of distributed generation.

The fuel market is the first area we should examine from this viewpoint. The market for fuels may require only modest adaptation to accommodate renewable energy sources, since most renewable fuels will provide a direct substitute. For example bio-diesel can operate in place of conventional or blended diesel, or bio-mass pellets in place of coal for co-firing in a coal power station. However, fuels that are not direct replacements – such as ethanol – will require consideration with respect to the market to ensure they are properly accommodated. This is of special concern because of the dominant position that vertically integrated fossil fuel companies have in the fuel supply sector.

A second important aspect of market reform is electricity systems. They differ widely between nations, from fully state-owned monopolies, to fully unbundled[7] and deregulated markets which have only private sector participants. Whatever the policy framework, it is likely that it has been built on, or derives from, a system of large centralized power generation. The small, modular, distributed nature of renewables requires major changes to the philosophy of such systems.

The following subsections will concentrate on electricity, but where appropriate we may refer to fuels.

Are the certification or licensing requirements appropriate for renewables?

Pricing, requirements and responsibilities must reflect the small but modular nature of the renewable generation plants. Renewable project sizes may vary from several hundred megawatts down to a few megawatts or even a few hundred watts of rooftop PV. Therefore, licensing intended for coal power stations or any other type of large, centralized generation is clearly not suitable. Similarly, fuels such as bio-diesel which are derived from crops may be more appropriately made and distributed at a local level, rendering unviable national certification or licensing designed for multinational oil companies.

In the longer term whatever replaces the existing system should take the form of a new licensing system rather than ad hoc waivers, while being mindful of the need for stability. Waivers imply interim fixes and occur at the whim of government, and there are business people who would consider some governments to be very whimsical indeed. Therefore the solutions must be long term and properly integrated into the existing licensing regime.

Is access to commodity markets affordable and guaranteed?

Renewable plants either produce electricity or fuels as their commodities. If markets are found beyond the home or farm, access to them needs to be both physical and legal. Producers must be able to get their commodity to buyers without hindrance, for example, through power lines, and they must have the ability to legally sell the commodity in the market place.

Access to distribution is primarily an issue for renewable electricity (assuming here that fuel producers have the greater ability to control their own distribution).

As the case studies in this book illustrate, access to grid hardware and systems has proved to be a huge stumbling block for renewable generation in almost every new market, and must therefore be examined very carefully.

Problems normally occur when the entities responsible for connecting renewable generators have too much discretion or inadequate guidance regarding the cost and hardware requirements for connection. They may also have conflicts of interest through equity holdings in competitors or potentially displaced generating assets.

By access to electricity systems, I am referring to issues such as licensing, connection contracts and fees, system requirements for safe integration of the power supplied, costs for moving electricity through the network, and taxes and fees. A final issue is fair recompense for renewable plants that reduce costs for the distribution company through reduced importation of energy into the local network or by offsetting the need for network upgrades. This is not as relevant for intermittent supplies where the transmission back-up is still required, but is more applicable to constant or dispatchable renewables such as small hydroelectric facilities or biomass plants.

The distributed nature of renewables means that the large production volumes are in fact amassed from many small projects. This makes access negotiated on a case-by-case basis inefficient and inappropriate. Although a one-off negotiation might be appropriate for a large power station, for renewables the access for many small, similar projects must be standardized and streamlined. Repeated negotiation for access creates unnecessary work and causes excessive transaction costs for small projects, and this results in competitive distortions between developers with different access costs.

Is there prioritized dispatch for renewables that cannot control their production times or volumes?

Some renewable energy projects, such as hydro schemes and biomass plants, can control when they dispatch energy or have quite stable production levels. Others, such as wind and solar, produce energy as it is provided by the natural forces harnessed. This poses special challenges for renewables like grid-connected wind and solar unless there is utility-scale power storage.

Those renewable projects that are able to control their delivery can maximize the value of their energy by delivering during peak demands and at peak prices in the system. Those unable to choose when they dispatch require a system to ensure their production always has take-up in the grid and is not denied access. This is somewhat similar to the needs and control issues of large thermal stations such as coal and nuclear plants, in that they cannot change their output as rapidly as demands change.

It is important to clarify here that the needs of one individual renewable project should not be seen to limit the entire industry or resource. Wind power may be considered a fluctuating resource, but total wind production across entire, large grids provides much steadier supply and in fact approaches base-load characteristics. Although production in one locale may vary, total production over many locations aggregates into something much more stable. This is because while one

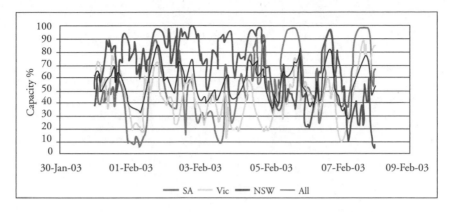

Source: Coates (2003)

Figure 3.14 *Example of smoother mean total output when multiple, geographically diverse wind sources are combined*

wind farm may experience a lull, it is highly unlikely that distant wind farms are also experiencing a lull. In fact, the physics of the atmosphere pretty much preclude this. This tendency of production from widely distributed wind farms to show smoothed characteristics is illustrated in Figure 3.14.

Obviously each renewable has its own set of operating conditions. Solar energy is produced only during the day, but an advantage is found in hot countries where solar PV output closely matches demand peaks caused by use of air-conditioning. Biomass is dispatchable and storable, but may tend to be quite seasonal in terms of production volumes. Small hydro will be seasonal and vary with precipitation levels from year to year, but can sometimes be stored for several hours to allow for response to daily peaks. Wind is intermittent with pressure fronts, can be diurnal but can also exhibit stable patterns such as found with trade winds. Geothermal can provide stable base-load power and its supply can also be increased or decreased for limited periods, although this leads to a loss of efficiency.

Is there transparent pricing throughout the grid to allow fair prices for renewables?

Large renewable installations are typically located in rural areas where the cost of delivering electricity is expensive and transmission losses are substantial. Users in these areas often have their electricity prices cross-subsidized by urban users.

Renewable energy can reduce these costs by putting energy production closer to where it is used and thus deferring network upgrades and reducing transmission losses. To actually allow this to happen, the relative costs of electricity throughout the grid need to be known. These costs will include the price of generation, the cost of infrastructure to get the power there, the cost of losses, and any additional taxation.

Once this price is published for all areas, the offset costs of embedded generation can be quantified. These are the savings produced by the locally established

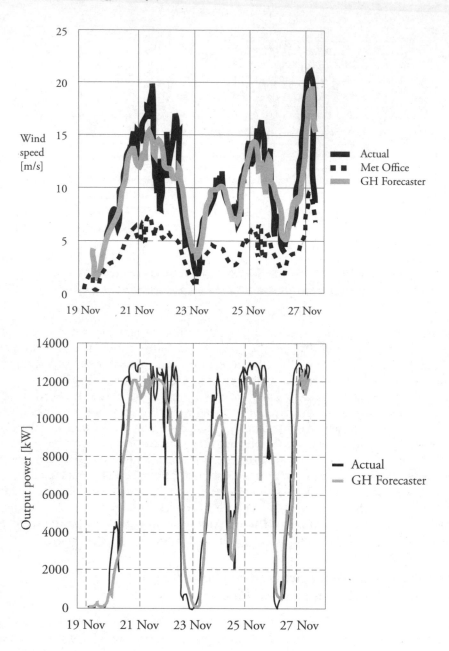

Note: While some renewable energy sources are intermittent, this does not mean they are unpredictable. Forecasting technology that advises approximate production a day in advance provides ample guidance for many markets. GH = Garrad Hassan
Source: White (2004)

Figure 3.15 *Wind forecasting twelve hours in advance versus actual wind and actual output of a wind farm*

renewable generation facility, savings which can then be fairly transferred from the distribution company to the generator. This transfer not only increases the viability of renewable energy projects, it also leads to more efficient outcomes for the distribution company and network planners. It is a potential win–win situation for everyone – provided that it is made to happen.

Does the framework provide supportive cost distributions for infrastructure changes or upgrades?

Major changes to national energy infrastructure can unlock vast renewable resources, but how are these costs managed? Such infrastructure changes should not be left to the free market or a handful of developers. This infrastructural evolution reflects a strategic pathway to unlock national resources and must be treated accordingly.

The precedents set by thermal generation are appropriate here too. The approach of distributing cost over the entire consumer base may be quite applicable to renewables, just as it is typical for large conventional plants in many countries. The new transmission is then considered a national infrastructure asset. To ask renewable developers to pay for major new infrastructure when existing thermal generators have had all such costs defrayed by consumers or taxpayers would be anti-competitive and create a significant market distortion.

The infrastructure upgrades required can range from the very local, such as increasing the capacity of line to a few farms, to the very large such as laying cables out to wind farm locations in the North Sea. There are perhaps two equitable ways to distribute this cost. The first is to spread the cost across the full consumer base, as is done for large infrastructure upgrades for power stations. The other is to have the taxpayer absorb the investment through an enabling programme.

Land Use Planning Reform

While the issue of land planning is a sub-set of the policy frameworks point above, this aspect is nonetheless drawn out to receive further consideration here. This is because of the distributed nature of renewable energy projects, the need for many small projects, and ongoing evidence that planning obstacles can cause major delays for renewable development.

Later, this book will show how many stakeholders are affected by renewable energy development. Stakeholders range from residents and neighbours whose communities host renewable energy developments, to abstract global populations whose future wellbeing will be safeguarded by rapid action to reduce greenhouse gas emissions. Stakeholders may have an economic, social or environmental interest in a project. They may be local or non-local. They may be directly affected or may advocate on behalf of others. Many stakeholders will come into direct contact with projects through the planning application process, thus it is important that planning policies are established to properly manage this interaction.

The UK, for example, provides a cautionary tale. Despite the highest wind resource in the EU15 and one of Europe's earliest promotion policies, ongoing obstacles have left developers and stakeholders without a properly integrated planning framework, as detailed in the UK case study by Gordon Edge in Chapter 6.

The UK example contrasts with that of Denmark, Germany and Spain where more integrated consultation and planning frameworks evolved. The following excerpt sheds light on the Spanish experience of environmental planning:

> *Environmental concerns [in Spain] have been given a different emphasis in different regions. Navarra included environmental impacts as one of the key aspects in site selection at the start. Other provinces such as Galicia and Castilla have not fully dealt with these issues leading to conflicts with organizations and residents. Other regions such as Catalonia have seen their plans delayed whilst awaiting proper decisions on how to address these conflicts.* (EWEA, 2002)

There are lessons to be learned from these experiences to help avoid such pitfalls in the future, and they are addressed in the following with some key questions and answers.

Are renewable resource maps available that can provide combined technical, environmental and social overlays to allow informed decision-making?

The first step in good planning is to ensure that sufficient information exists. Specifically, we need to know what resources the country has, in what volumes and where they are located. We also need to know what the infrastructure concerns are in terms of how it will affect the harnessing of these resources. For example, are roads required to bring fuel in? Are major power lines required to get power out? And is this infrastructure adequately mapped?

This must be overlaid with information about potential points of environmental impacts. For example, the locations of sensitive or special biodiversity must be recognized. Furthermore, we need to understand which types of impacts from the technology concerned must be considered in detail.

Finally we must understand possible social impacts. Considerations include population distribution, the impacts relevant to populations (such as noise, transport, building changes or fuel replacement), optimizing employment creation with location, positive or negative overlaps with other land use (e.g. farming), and issues of landscape sensitivity (see Box 3.4).

Much of the information required to answer the questions above will be readily available to a government. Sometimes, however, government may not have the information or the agencies that can do the work, or it may choose not to take the lead. In these situations civil society and the renewables industry must step into the gap. For example, the landscape arena is one aspect that may be left in the breach. Box 3.5 provides some guidance as to how a coordinated process might generate a landscape information base or methodology for some renewables.

Box 3.4 *Example of overlay use by SEAV in Victoria, Australia*

It is very important not to pre-judge outcomes of new developments or assume that all impacts will be negative; the evidence so far indicates the opposite for many renewable energy technologies. For example, wind farms have been shown to increase property prices in local areas (Sterzinger et al 2003). The more informed all stakeholders are, the more informed the debate and the decision-making.

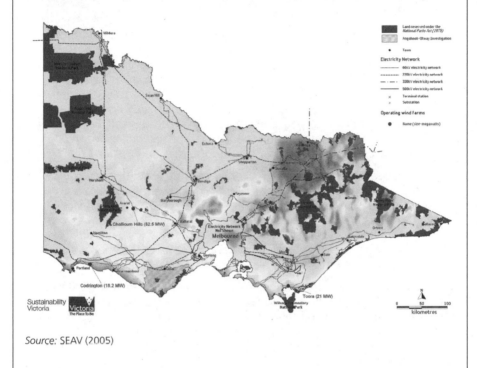

Source: SEAV (2005)

Figure 3.16 *Electricity grid, parks and wind resource overlays for wind development in Victoria, Australia*

Are there environmental and social impact standards in place to provide guidance to developers and security to stakeholders?

If we now assume that all the information is in place to make good decisions, the next step is to establish standards or planning requirements to ensure the impacts of renewable energy development are acceptable.

It is impossible to cover all impacts here. However, some of the common impacts include noise levels from wind farms at the nearest dwellings, impacts of wind farms on birds or bats, impacts of biomass crops on local biodiversity, impacts of biomass residue removal on soil quality, effects of transport levels on local roads and communities, impacts during construction, and decommissioning arrangements. For many of these impacts planning standards may already be

---

Box 3.5 *Australian Wind Energy Association (AusWEA) &*
Australian Council of National Trusts (ACNT) landscape protocols

This example outlines an initiative to improve information and policy on landscape and heritage. Here the Australian wind industry voluntarily worked with
a major stakeholder group to develop material for use by the industry. The following are excerpts from statements that the two organizations released to the
media in 2003:

> *The Australian Wind Energy Association (AusWEA) and the Austral-*
> *ian Council of National Trusts have announced an agreement to work*
> *together to ensure landscape protection as the wind industry grows.*
> *The two organizations have launched a joint project that will determine*
> *and promote an agreed-upon means for assessing landscape values*
> *inorder to ensure that the planning and siting of wind energy devel-*
> *opments can proceed, and that significant landscapes can be identi-*
> *fied. The federal government is supporting the initiative with funding*
> *to engage independent experts to advise the two organisations.*

> *'As the peak community organization concerned with Australian land-*
> *scape conservation, we are seriously concerned about the effects of cli-*
> *mate change on Australia's environment, but also the visual impact of*
> *climate solutions such as wind farming. We have come together with*
> *the wind industry because we both believe that a sustainable solution*
> *can be found in the siting of wind developments in the countryside,'*
> *said Mr Molesworth* [Simon Molesworth, ACNT national chairman].

The director of community relations for AusWEA, said: 'Our partnership with
the National Trust will allow us to work together across Australia to provide
local stakeholders and wind farm developers alike with better understanding
and confidence that their interests are being considered. As the expert groups
in wind and landscape we can also help inform the various government agencies on viable solutions.'

---

in place, but if they are not it is not uncommon to borrow standards from other
countries. However, some impacts will be country specific, and here forward planning can prevent mistakes or periods of uncertainty.

Again government may not always be the best entity to take the lead on some
of these issues, in which case industry and civil society may be left to play lead
roles.

Is there zoning or strategic mapping that minimizes planning risk for developers, and also minimizes confrontation between developers and stakeholders?

As previously stated, planning problems have caused difficulties for renewable
developments in many countries. The need for a large number of planning con-
sents is inherent in technologies that are small, numerous and widely distributed.

Therefore the more streamlined the planning process, the better for all concerned. Pre-empting these processes by adopting zoning or strategic maps to guide and streamline decision-making are useful ways to address this issue.

Experience shows that it is not wise to wait until planning problems emerge before seeking the solution. For example, proceeding with the process of developing a wind farm until it receives significant planning objection may serve to undermine an entire renewable energy policy drive. Often, objecting parties will not confine themselves to comment within the planning process, but will expand complaints to the media and politics too – causing reputational damage to both government and industry.

Pre-emptive planning action also makes it easier for governments to relate the overarching sustainability policies to planning reform. Should an objection exist, it is preferable to handle it within a balanced 'national–regional–local' planning reform or zoning initiative rather than a more polarized local debate where issues of climate change or national energy security may have little or no resonance.

Finally, there is strong evidence that the pre-existing attitude of local people to a particular development company will affect their attitude to a project owned by that company. Renewable energy in general should not be made hostage to fortune in every project area as a result of how some stakeholders regard the local developer. This risk can largely be removed if developers are constrained to develop appropriately, based on predetermined zoning or mapping or protocols, and in keeping with decisions already made by various levels of government. The following quote provides a Danish example:

> *Around 1997 another set of planning regulations were developed for offshore wind farms, with a central, national authority – the Danish Energy Agency – being responsible to hear all the interested parties, public and private. This 'one-stop shopping' method has facilitated the planning process considerably, and is widely studied around the globe.* (Krohn, 2002)

Equalizing the Community Risk Cost–Benefit Distribution

The environmental and energy benefits of renewables mainly accrue at a national and international level. The environmental and social direct impacts occur mainly at a local level. This discrepancy can lead to inequitable distribution of costs and benefits if they are not considered carefully.

The destructive impacts of climate change act at a global level, as do the positive impacts of greenhouse gas mitigation via renewable energies. For nations embarking on emissions reduction programmes or which are signatory to the Kyoto Protocol, the benefits of the renewable energy could be said to accrue at a national level. However, localized impacts or lifestyle adjustments will occur at the community level where the renewable projects are hosted. Since these impacts can be both positive and negative, it is important to ensure that local communities get their fair share of the benefits of renewable energy development so that they may consider themselves net beneficiaries.

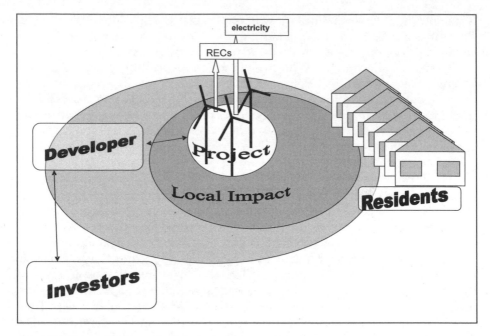

Note: RECs: Renewable Energy Credits.

Figure 3.17 *Many environmental and social direct impacts of renewable energy projects will be experienced at the local level, although the benefits of ownership and investment may not be offered to the local community*

Compelling rationales exist for government to intervene on behalf of the nation and find mechanisms to increase the benefit to local communities. Evidence from successful markets indicates that finding suitable avenues to increase the direct benefits to host communities dramatically increases their acceptance of the projects and willingness to host further renewable energy developments. The following questions and answers could help to guide such outcomes.

Are the local impacts of developing renewable energy projects of this form known and quantified?

The local community impacts may be negligible for renewable installations such as solar photovoltaics or solar hot water heating. However, even these will need factories and raw materials. Where will these factories be built? Where will supplies be sourced? Who will this impact? Other technologies such as wind, geothermal and biomass may result in significant disturbance during the construction phases and entail some level of impacts thereafter. As previously mentioned, typical effects might include: visual amenity and changes to the landscape, increased traffic from fuel deliveries, and land use change for energy crops.

The process of identifying all the impacts for the various renewables is reasonably simple and has been done for countless projects, often in the form of environmental impact statements. With this information in hand it is possible to look at how balanced this equation is for host communities and other stakeholders.

Do the policies require that communities and stakeholders are properly informed and consulted, and the potentially negative impacts thereby minimized?

Before considering the possible solutions, it is important to appreciate that the process itself often colours stakeholders' view as to whether an outcome is balanced or not. Three steps that can really assist in this regard are: information, consultation and adaptation.

Providing insufficient information is a classic and repeated failure of governments, industries and non-governmental organizations (NGOs) alike. Renewable energy industry development often starts in an information vacuum. Information is often only developed in response to community or stakeholder concern and it may well be that by the time correct information is made available the vacuum has already been filled with rumours, half-truths, untruths and scandal.

Almost every person on the planet has an innate fear of the unknown. It is an inescapable human characteristic. Renewable energy projects create changes and introduce unknowns. A crucial service that must be performed is to minimize the unknowns through the provision of balanced facts and experience. People feel respected and appreciated if they are kept informed.

Consultation is also crucial. Personal acceptance of imposed change is affected by how much that person or group has been consulted and listened to. This is over and above the degree to which their opinion has prevailed. People understand that most outcomes will require a balance on issues and opinions, but they also feel it important that their voice be heard and acknowledged in the process.

Specialist stakeholders will also provide a valuable source of expertise on environmental and social issues. However, one aspect of consulting such groups is that their resources may be limited and this constraint needs to be accommodated.

Next comes adaptation; consultation should be followed by a process in which opinions, concerns and proposals are considered and projects adapted where necessary. Sometimes requested changes may make little or no impact to a project's viability. On the other hand, more rapid approval can save significant amounts of time and money.

There is considerable merit in the incorporation of formal government or industry policies to ensure that best practice is followed for information, consultation and adaptation. Good renewable energy developers will follow best practice anyway, so for them this will not be an imposition. However, the policy will serve to pick up the more lazy developers and avoid community confrontations that have a damaging effect on the wider renewable energy industry.

Box 3.6 *Using television to inform the public about wind energy*

What do wind farms look like? Is it better for the public to read an academic argument in the newspapers or see wind farms themselves? In 2003 a television advertisement was screened in various parts of regional Australia on behalf of the AusWEA. The two stated objectives were to demonstrate why renewable energy is needed and to show people what wind farms look like in moving images. While the use of television might not seem to be the most obvious form of public outreach for wind developers, it is an effective way of letting a very influential stakeholder group (the public) see this technology in action. The ad resulted in greater support for the MRET campaign through the Internet, while the television station that aired the ad received calls from the public expressing positive feedback.

Source: Transition Institute (2003)

Figure 3.18 *Screenshots of Transition Institute advertisement for the Australian Wind Energy Association*

Have the potential positive impacts for local and regional communities been maximized within the policy framework?

It is important to maximize local benefits to ensure that the positive and negative impacts of a renewable energy development are at the very least balanced, and at best weighted in favour of the local community and stakeholders. Much of this may be up to the project developer, but because we are looking at policy initiatives, we will examine some of the more universal techniques that can be applied.

We need to be very aware that these measures may have some cost impact on renewable energy; if they were the cheapest option they would likely happen anyway. However, policy intervention is necessary to ensure that the measures to maximize local benefit occur, and this can be done in one of two ways: with the carrot or the stick. Whether we use incentives or obligations to achieve the desired outcome depends on which is most efficient in a given situation.

There are numerous potential measures. They include creating import tariffs or tax breaks that provide an incentive for local manufacture, or requiring specified levels of local content or spending in planning approvals. Incentives can also be provided to maximize the amount of locally spent capital investment. Provision of incentives or investment is also possible for re-training and company start-up to maximize the amount of installation, operation and maintenance spent locally. Measures may also include mechanisms for local ownership, and incentives that increase benefits for local investors over non-local investors.

We will consider local employment and local ownership in a little more detail as we answer the next two questions.

Does the policy framework provide incentives for local employment content?

For every person directly employed by a company, typically another two are indirectly employed. Employment creation can have a dramatic impact on small local economies.

Renewable energy sources are especially labour intensive. Leveraging job creation at a local level significantly increases the actual positive social impact of projects. Many of the policies and measures discussed above contribute towards this outcome. However, additional incentives can be put in place.

Renewable development can create employment in project development, engineering and consulting, construction, installation, fuel collection, and operation and maintenance. In fact, employment in some of these job categories can be very significant. For example, a 1.5MW wind farm may require spending of about US$25,000 per year for operation and maintenance (a 50MW wind farm $750,000 per year) and much of this employment can be provided locally.

India has used import tariffs very successfully to maximize local manufacture of content for renewable energy production. Moreover, the government has struck a clever balance between stimulating Indian content without increasing prices. In 2001, I led a team to Gujarat, India following a major earthquake there, to install emergency stand-alone power equipment in schools and hostels. We sourced as much as possible from local companies. One such company explained that they were able to import from overseas individual solar cells with little tariff impost, as these components were not manufactured in India. These cells were then used to locally assemble the complete solar panels. Whole solar panels, on the other hand, had a significant import duty since this local manufacture was possible. This policy clearly provides an incentive and advantage to locally assembled products.

Techniques such as these provide incentives without burdening government or taxpayers. Indeed stimulating economic activity and employment will have a net positive effect for government in most instances.

Are there policies that provide means and incentives for local ownership?

There is strong evidence that local financial ownership can have a dramatic effect on acceptance. The Danish have had particular success linking tax benefits for renewable energy investment to geographical proximity of projects. Local people are thereby singled out for special benefits from a nearby project, as the following quote from Dambourg and Krohn (1998) shows:

> In Denmark 80 per cent of erected turbines are owned by individuals or co-operatives. More than 150,000 families have shares in wind energy schemes. The connection between ownership and social acceptance is set out by Krohn (1998).
>
> The highest concentration of wind turbines in the world occurs in a place called Sydthy, Denmark. Sydthy has 12,000 inhabitants and 98 per cent equivalent of its power comes from wind power. The reason that the community accepts and allows so much wind development may be explained by a poll by Andersen et al (1997), that reveals that 58 per cent of the households in the municipality of Sydthy have one or more shares in cooperatively owned wind turbines.
>
> The Danish Wind Industry Association overview of international public surveys states, 'In Denmark there is a tradition for wind co-operatives, where a group of people share a wind power plant... Regarding the general attitude towards wind turbines, the picture is clear. People who own shares in a turbine are significantly more positive about wind power than people having no economic interest in the subject.'

In the early days of Danish wind development, wind turbines were smaller and therefore within the financial reach of farmers and co-operatives. As new renewable projects start to increase in size and value to well over $50 million, they are beyond the grasp of the home-spun owner or collective. In new markets this presents a novel challenge.

Intervention will be required to create the means for local ownership in these larger renewable energy projects. This intervention is necessary because, first, the scale makes the contribution of individuals minor; second, there is no shortage of global investment capital meaning local investment is not required; and finally because the cost of servicing many small investors is greater than the cost of dealing with a single large institutional investor.

While it is beyond the scope of discussions here to explore policy intervention capable of solving this ownership hurdle,[8] examples do exist. One is the Middlegrun offshore wind project near Copenhagen, a development that is 50 per cent utility owned and 50 per cent shareholder owned.

Conclusions

In this chapter we have attempted to cover the basis of a sound renewable energy policy framework. In a way we can think of the issues and the questions that policy-makers need to consider as forming a checklist for sound policy. If this book were to be condensed into one page, perhaps this checklist would be it:

1   Transparency:
    - Are the policies comprehensible enough to understand and comprehensive enough to cover all of the components required to make projects bankable?
    - What unknowns exist in the policies that might affect the market size, the prices paid for renewable energy or the duration of the scheme?
    - Are the policies structured fairly so that they do not favour insiders compared to outside entrants?
    - Do the policies have sufficient time frames to allow dissemination and engagement from interested parties in other countries and sectors?
    - Are the renewable energy policies consistent with policies and measures for other parts of the energy sector, or are there mixed messages and double standards?
2   Well-defined objectives:
    - What outcomes are actually intended from the renewable energy policies?
    - Are the drivers and measures specific about the intended outcomes?
    - How is performance checked and the risk of under-performance minimized?
3   Well-defined resources and technologies:
    - Do the policies avoid putting industries which are extremely different in size and maturity into direct competition?
    - Are the policies focused on renewable energy free of contradictory goals?
    - Are technologies treated differently if necessary? And how?
    - Is the intended technology or mix of technologies properly articulated?
    - Are the consequences of including any environmentally unsustainable technologies fully understood?
4   Appropriately applied incentives:
    - Is there flexibility to allow new technologies to be included and to evolve in future?
    - Are the identified technologies actually being developed by the policies?
    - Is the line drawn around the eligible technologies in a way that allows adequate promotion for the country-appropriate technologies?
5   Adequacy:
    - Are the policies leveraging private sector investment?
    - Are the returns on investment comparable with other alternatives?

- Are the investment periods long enough for project investors to get a return on equity?
- Are the installation intensity and scheme duration adequate to enable manufacturing?

6 Stability:
- Does the policy framework avoid boom–bust cycles?
- Does the policy plan for the whole cycle of an industry's development?
- Does the policy provide an ongoing steady pull on development?
- Is the resource base for the incentive sustainable?
- Are there long-term energy policies that provide guidance and surety for evolving policies and measures?

7 Contextual frameworks:
- Is there an overarching national policy objective in place to guide all policy-making?
- Have all the laws that affect renewable energy projects been checked to ensure they are conducive and not obstructive?
- Are the policies and measures at all tiers of government consistent and self-reinforcing?

8 Energy market reform:
- Are the certification or licensing requirements appropriate for renewables?
- Is access to commodity markets affordable and guaranteed?
- Is there prioritized dispatch for renewables that cannot control their production times or volumes?
- Is there transparent pricing throughout the grid to allow fair prices for renewables?
- Does the framework provide supportive cost distributions for infrastructure changes or upgrades?

9 Land use planning reform:
- Are renewable resource maps available that can provide combined technical, environmental and social overlays to allow informed decision-making?
- Are there environmental and social impact standards in place to provide guidance to developers and security to stakeholders?
- Is there zoning or strategic mapping that minimizes planning risk for developers, and also minimizes confrontation between developers and stakeholders?

10 Equalizing the community risk/cost–benefit distribution:
- Are the local impacts of developing renewable energy projects of this form known and quantified?
- Do the policies require that communities and stakeholders are properly informed and consulted, and the potentially negative impacts thereby minimized?
- Have the potential positive impacts for local and regional communities been maximized within the policy framework?
- Does the policy framework provide incentives for local employment content?
- Are there policies that provide means and incentives for local ownership?

Notes

1   This does not mean that price support or other mechanisms are unending as all project support can still have finite timelines. Rather it means that the legislation or application rounds do not have restrictive dates which lead to boom–bust cycles and unstable pricing.
2   Producers may be liable for the full product life cycle; for example, battery companies may be required to take old batteries back when selling new ones to ensure safe disposal or recycling. Similarly it has been proposed that the cost of carbon pollution damage could be applied to fossil fuels sales via extended producer liability.
3   With regard to greenhouse gas emissions in particular, the World Commission on Dams (WCD, 2000) points out that hydroelectric installations may not always be a net benefit due to the conversion of soil and plant to carbon in methane after flooding: 'All large dams and natural lakes in the boreal and tropical regions that have been measured emit greenhouse gases … some values for gross emissions are extremely low, and may be ten times less than the thermal option. Yet in some cases the gross emissions can be considerable, and possibly greater than the thermal alternatives.'
4   'Such improvements to overall system efficiency have been achieved by greater aerodynamic efficiency of the rotor blades, the use of high-efficiency electric conversion systems and better matching of the wind turbine rating to the local wind regime' (IEA, 2003). Availability, that is the fraction of time a wind turbine is available to generate energy rather than, say, off-line for servicing, is now approaching 98–99 per cent for modern turbines and is considered to have reached its limits.
5   Models and tools to analyse the steady and dynamic impact and operational characteristics of large wind farms, improved wind forecasting and development of various enabling technologies will increase the value of wind power (NREL, 2003).
6   If something is expensive, is not the obvious solution to throw money at the problem? The fantastic verb, 'to apollo', comes from the US Apollo space mission, into which the government funnelled a wide and seemingly endless pipeline of cash to crash through the technical obstacles of a moon mission. Thankfully, the renewable energy industry does not yet require 'apollo-ing'. The only period of such intensive spending was after the 1973 oil crisis, when companies like Boeing and Westinghouse applied their expertise to the wind turbine challenge. This resulted in only a single surviving US wind turbine manufacturing company (now owned by GE).
7   Unbundled means that the generation, transmission, distribution and retail sectors of the industry have been separated to allow competition and avoid control by monopoly holders.
8   This is the subject of a Transition Institute business incubation project, the Community Wind Trust. See www.transitioninstitute.org

References

BTM (2004) *World Market Update for 2003*, BTM Consult ApS, Ringkøbing, Denmark
Charles River and Associates (CRA) (2003) *Economy Wide Effects of Increases to the Mandatory Renewable Energy Target*, Charles River and Associates report to the Renewable Energy Generators of Australia, June
Coates, S (2003) 'Maximising the benefit of wind', presentation to the SEDA Renewable Energy Seminar, 6 November, Sydney
Dambourg, S. and Krohn, S. (1998) *Public Attitudes Towards Wind Power*, Danish Wind Industry Association, Copenhagen
DTI (2003) *Energy White Paper: Our Energy Future – Creating a Low Carbon Economy. Version 11*, UK Government Department of Trade and Industry, published February, www.dti.gov.uk/energy/whitepaper/wp_text.pdf.
European Wind Energy Association (EWEA) (2002) *Windforce 12*, European Wind Energy Association and Greenpeace International, Brussels
Grubler, A., Nakicenovic, N. and Victor, D. G. (1990) 'Dynamics of Energy Technologies and Global Change', *Energy Policy*, vol 27, pp247–280

International Energy Agency (IEA) (2002) *Renewables Information 2002 Edition*, International Energy Agency (Statistics), Paris

IEA (2003) *Renewables for Power Generation: Status and Prospects*, International Energy Agency, Paris

Krohn, S. (2002) *Wind Energy Policy in Denmark: 25 years of Success – What Now?* Danish Wind Industry Association, Copenhagen

Maddox, R. (2004) *What Does Wind Energy Mean to You?* Private Communication, July

Mallon, K. and Reardon, J. (2004) *Cost Convergence of Wind Power and Conventional Generation in Australia*, Version 1.34, June, Australian Wind Energy Association

McDonald, A. (2004) *Nuclear Expansion: Projections and Obstacles in the Far East and South Asia*, World Nuclear Association Annual Symposium, London, September

National Renewable Energy Laboratory (NREL) (2003) *Power Technologies Data Book*, Energy Analysis Office of the National Renewable Energy Laboratory, Golden, CO

SEAV (2005) *Victorian Wind Atlas*, Sustainable Energy Authority Victoria, Melbourne, Australia

Segurado, P. (2005) *React Renewable Energy Action, National Report, Spain IDAE*, Produced for EU ALTNER National Meetings and Lessons Learned, 23 March

Sterzinger, G., Beck, F. and Kostiuk, D. (2003) *The Effect of Wind Development on Local Property Values*, Renewable Energy Policy Project, Washington, DC

Transition Institute (2003) *Or We Can Do This*, screenshots from television advertisements produced by Matt Trapnell for AusWEA, downloadable from www.transitioninstitute.org

World Commission on Dams (WCD) (2000) *Dams and Development – A New Framework for Decision Making*, report for the World Commission on Dams, Earthscan, London, released November 2000 by the WCD, www.dams.org

White, G. (2004). *Wind Forecaster Presentation*, produced by Garrad Hassan Pacific, private communication

4

An A to Z of Stakeholders

Karl Mallon

In the grand task of winning heart and minds,
one first needs to know who owns them.

Stakeholders are the central part of winning policy change of any kind. This has never been more true than for renewable energy. Renewables entering a new market are confronted with a highly competitive energy sector. They are offering an expensive product in a sphere already populated by powerful players. Some of the benefits of renewable products are abstract and long term. If ever anyone needed friends in the right places, it is a new renewable energy industry.

The aim of this chapter is to construct a framework that allows us to identify and understand parties that are stakeholders in the process of renewable energy development. We will do this by examining why stakeholders are so important, by defining stakeholders themselves, and then by defining the effects of renewable projects on these stakeholders. Next we will develop a stakeholder matrix, a useful exercise for any party seeking to recognize and analyse stakeholders. We will also look at the issue of stakeholder resources, an important consideration for measuring up potential proponents and opponents of renewable energy development.

In the second half of the chapter, we will use this analysis to identify stakeholder risk management and engagement strategies that can maximize positive and minimize negative outcomes. Finally, we will take a close look at some key stakeholders in terms of their likely stance towards renewable energy development.

Why are Stakeholders So Important?

Renewable energy development occurs only if the status quo is changed. This change may be happening anyway due to power shortages or environmental imperatives, or it may need to be driven. An embryonic renewable industry comprising a handful of individuals and businesses can do little on its own. It must leverage help from more powerful allies and constructively engage current and potential critics.

Box 4.1 *First exposure*

The ideal situation is that people receive information that is both engaging and factual, be that from a company, industry association or NGO. In the last chapter we mentioned the use of TV advertising. But it may be a poster or advert in the paper like the UK's 'Embrace Wind' campaign.

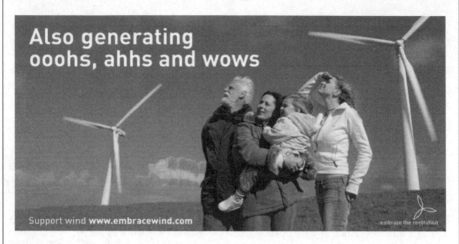

Source: BWEA

Figure 4.1 *The billboard poster for the British Wind Energy Association promoted higher renewable energy targets while also conveying the benefits of renewable wind power to the general public*

Compare this as a first exposure with hearing a radio show like the one extracted below, with comments by a popular British botanist who now flies around the world to talk as an expert on wind farms:

> *Famed environmental campaigner Professor David Bellamy yesterday backed Noel Edmonds' anti-wind farm campaign, describing them as a 'scam'.*
>
> *The botanist, writer and broadcaster is a staunch opponent of wind turbines arguing that they are inefficient, destroy the landscape and that far more could be achieved through energy efficiency.*
>
> *'My main thing against them is that they can only work, if you are very lucky, for 30 per cent of the time,' he said yesterday. 'Going by the ones in Denmark it is about 17 per cent of the time.*
>
> *'So how are people going to be able to boil their kettles, or how are we going to power our hospitals the rest of the time? It means that we have got to keep our other stations running, spinning in reserve, inefficiently and pouring out carbon dioxide and sulphur dioxide and the like.'* (Country Guardian, 2003)

In practice, renewable energy will be on the agenda in a given country because some pro-renewable stakeholders have put it there: they may be interested academics, environmentalists, energy providers or others. Such stakeholders will be active even before the renewable industry itself.

Equally, renewables will have a set of impacted stakeholders with which it must contend. These people and organizations will have a stake due to the effects renewable energy development will have on the energy sector specifically, the economy more widely and the physical placement of actual projects in environmental and social space.

Many of these parties will not actually realize they are stakeholders at the outset of renewable energy development. Their reaction to renewable energy upon realization that it affects them will often be coloured by the party that contacts them first and the information they receive. If the first news a local resident receives about a wind development is that turbines sound like Second World War bombers and look like a skyscraper with no windows, that is not a promising start. If by contrast residential energy consumers are engaged in the vision of a clean energy future, the harnessing of a local resource and a new rural industry, they will often be supportive.

Apart from the politically important role stakeholders play in the contest to win a place for renewables in energy markets, stakeholders also play a useful role in the actual development of renewable energy. They help identify all aspects that must be considered to achieve an industry that provides maximum possible benefit across all sectors of society.

Some Ways to Define Stakeholders

To help define stakeholders, it is useful to ask who they are and what interests they have or represent. This section answers the 'who' question, while the next section will examine their interests.

The obvious way to define stakeholder types is to employ definitions that stakeholders will often use to identify themselves. These are citizens/individuals if they are acting alone, companies if they are commercial stakeholders, NGOs which can cover both commercial and non-commercial interest groups, and finally government at all its levels.

Citizens

People will in some way be affected by renewable energy development, from those who enjoy the global environmental benefits down to people who experience the very localized project effects. When wishing to have their interests represented, individual stakeholders will often look to government representatives or use existing or specifically created non-governmental advocacy organizations.

In so doing, they can act with great influence locally, regionally or nationally by writing letters to politicians or the media, or engaging in public consultations such as planning processes.

Companies

An equivalent individual commercial entity is a company. Company and small business stakeholders will often be locally based and have significance at a local level. At the opposite end of the scale, corporations can be massive transnational entities able to access government at its highest levels. Indeed some, like BP, Shell and General Electric, are active industry participants in both the renewable and fossil fuel sectors.

It is not uncommon for significant national companies to become leaders in renewable energy. TATA is a huge Indian corporation which is also a leader in photovoltaics (PV) through an alliance with BP. Similarly Kyocera and Sharp, giants in the electronics industry, are now also global PV giants. This is unlikely to be a coincidence; it likely indicates these progressive and outward-looking companies are better placed to exploit the domestic opportunities that renewables provide.

Non-governmental organizations (NGOs)

As the name indicates, this term covers just about any form of organization that is not a government body. NGOs range from non-profit organizations through to environmental or other pressure groups and on to commercially oriented industry organizations. This category can also include non-issue-specific groups like rotary clubs or local civil society bodies. However, it is important to caution that occasionally what appears to be grassroots may in fact be 'astro turf': representation attributed to community organizations or NGOs but actually orchestrated by industry (if you want an eye opening read on astro-turf look up Beder, 1998).

Government

Governments and their agencies are not often identified as stakeholders. Yet, in addition to being the decision-makers, they are also the ultimate stakeholders. They are the stakeholder affected by how all the other stakeholders are affected. And ultimately they are the stakeholder that has to keep the whole show on the road – economically, environmentally and socially. From local councils all the way up to international negotiating teams, governments are decided stakeholders. (See Figure 3.13 for tiers of government policies, plans and legislation used to deliver or facilitate renewable projects.)

The media

This important stakeholder is not often mentioned even though it mediates the interaction between all stakeholders. The media is more than just a conduit of information; the media can represent the renewable energy industry as the David fighting the good fight, or as the Goliath imposing its might. It can glorify the brave new world of sustainability or champion half a dozen project objectors as though they were an entire social movement.

Each nation's media culture is different and so there are few general rules – except that the media's status as the main information conduit may make it the most important of all stakeholders. However, there are three further important points to note. First, the media tends to be drawn to confrontation. Second, the media often uses the principle of balance, essentially meaning reporters will usually seek an alternative view on any given issue. Third, the media is made up of people, some of whom may be stakeholders because of, say, landholdings or share holdings, and this may colour their view or editorial position.

Since the media is all places and with access to all levels, I will not attempt to include it under the stakeholder framework we are building. But we must still consider the media all-pervasive!

Some Ways to Define Effects on Stakeholders

The next step in building a framework with which to understand and appreciate stakeholders is to consider the sorts of impacts that stakeholders of the above types may experience or perceive. Here I attempt to set out four discrete impact types to help us differentiate stakeholders later on. These are: direct economic, indirect economic, environmental and social impacts.

Direct economic impacts

Some stakeholders will experience a direct economic effect from growth in renewable energy deployment. This category will mainly be populated by beneficiaries (those stakeholders who suffer dis-benefit are captured under the 'indirect economic impacts' category). Examples include project developers, financiers, employees, fuel suppliers and landowners, and local councils receiving rates.

Indirect economic impacts

Indirect economic stakeholders are financially affected by renewable projects' secondary impacts upon individuals, communities or even countries as a whole. These impacts could include local people who benefit from a general rise in commercial activity, industries which lose a market share to renewables, unions for renewable workers or competitor industries, consumers who have increases in their energy bills, and people whose property prices increase or decrease.

Environmental impacts

Local or global ecology, climate change and species-specific impacts are examples of issues covered in the realm of environmental stakeholders. Environmental impacts described herein will refer to direct physical changes to biological activity and diversity, as distinct from social impacts.

Social impacts

Social impacts are those that may affect human interaction in positive or negative ways, as distinct from economic and environmental concerns. These might include changes to a landscape, re-routing of a walking path or establishment of a turbine viewing tower. This category covers impacts where there is no direct biodiversity or economic impact, but rather an aesthetic or social change.

Other stakeholders

Other stakeholders that defy cataloguing within these four categories will no doubt exist, for instance the media. Other interesting examples include the military, who have been major participants in deliberations over North Sea offshore wind installation placements, for example. Transport agencies will be concerned with trucks carrying biomass, and aviation agencies dealing with high wind turbines. Television transmitting companies and electrical standards agencies will be concerned about the effects of distributed generation on power stability. Thankfully, many of these parties will simply want to ensure no interference or disruption to their services and are unlikely to be hostile to renewable energy development in general (although they can set out overly cautious and therefore obstructive technical standards).

Notwithstanding parties that do not quite fit the mould, we now have two dimensions with which to define our framework of renewable energy stakeholders – the stakeholder types and the issue types.

A Matrix of Stakeholders

Based on the foregoing discussion we can now start to build up a matrix for stakeholders based on their type and effect. An example is placed in each cell of Table 4.1.

Table 4.1 *Forming a matrix of stakeholder type versus effect, with an example for each intersection*

	Economic direct impacts	Economic indirect impacts	Environmental impacts	Social impacts
Citizen	Employees in industry Project contractors	Energy consumers Populations local to projects	Bird watchers	Local heritage enthusiasts
Company	RE manufacturers	Energy retailers	Farmers	Tourism companies
NGO	Tourist organizations	Unions	National environment groups	Hiking groups
Government	Local councils	Ministry of regional development	Environment ministries and agencies	Heritage, cultural and conservation agencies

Populating the stakeholder matrix

Now that we have the framework, the next job is to populate it. There are three useful ways to consider identifying stakeholders: going from first principles; using the media; and looking for self-identifying stakeholders.

Every country, region and technology will have its own particular stakeholders and I cannot hope to capture them all here. Nevertheless, it is often better to identify stakeholders before they identify themselves, because early engagement can avert undue concern and help create allies rather than enemies. So the more specific we can be about potential stakeholders here the better.

Stakeholders from first principles – The scope of impact of renewable energy projects

With a field that can be as open-ended as impacts, it is useful to start from first principles. It is a simple – but extended – exercise in stepping through what would happen if a renewable energy industry developed.

This is similar to going through the process of an environmental impacts assessment or a socio-economic impacts analysis. Indeed it may prove very useful to acquire such reports whether produced locally or from overseas to examine the impacts identified. If projects are built, will there be jobs created and will other industries lose jobs? If so, unions will be stakeholders. Are animals saved by protecting the climate? If so ecological groups are stakeholders. Will wind farms be noisy or will biomass plants cause traffic jams? If so, local residents will be stakeholders. Will the economy be boosted or brought to its knees? The government is always a stakeholder.

It is perhaps useful to keep in mind three levels of effect.

Local

This category covers the physical environment affected by the project, the social/shared spaces such as to scenic views or walking paths, and the local communities who will host projects, provide employees, or who may have concerns about their property values.

Regional

Some regions are endowed with particular renewable sources and may therefore see multiple projects moving into the area, with cumulative effects. These can be positive such as the creation of manufacturing capacity and significant employment, or potentially negative due to over-concentration or unplanned development.

Total

Many effects only accrue at a national or even international level. Cost impacts may be distributed across all national consumers or taxpayers. The emissions savings will benefit people all over the world.

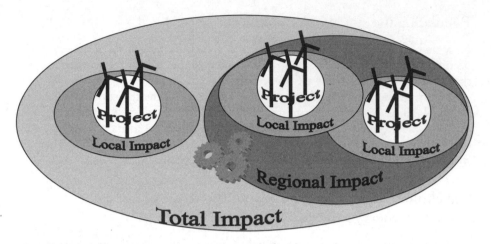

Figure 4.2 *The effects of renewable energy development can be considered local, regional and total and so, therefore, can the stakeholders*

Using the media to identify stakeholders

As opposed to considering the various impacts as a method to identify areas of stakeholder concern, it can be very enlightening to look at the issues that stakeholders themselves identify through media reportage.

This also provides industry players with insight into how they are being perceived. The actions a set of stakeholders take with respect to renewable energy will not be based on fact per se but rather on their understanding of that industry, project or technology. The best case is when their perception is close to fact, however, this is rare since few stakeholders become renewable energy experts. Instead, they will form their opinions from information they are exposed to, and this may not always be correct information.

The key sources of independent information are considered in most countries to be the press, radio and television. In fact this content may reflect limited access to expertise or even access to incorrect information, and may also be influenced by sensationalism.

Even though it may be distorted, media analysis can help to identify the vocal stakeholder groups. These are definitely people that the industry needs to engage and be seen to engage by those still considering the issues. It can also help to identify the groups that are remaining silent.

The very simple media analysis used in Figure 4.3 highlights the difference between impacts and perceived impacts. Media monitoring companies can provide such analysis reports for a fee, allowing an independent view of how an industry or technology is being conveyed to the public.

Figure 4.3 is an analysis of print media coverage of wind power issues in Australia from early 2000 through to mid-2002. The articles covered local, regional and national print media as compiled by the company Media Monitors.

There are some caveats on how much to infer from such an exercise. First, if the coverage is cited as discussing an objection it does not imply either a positive

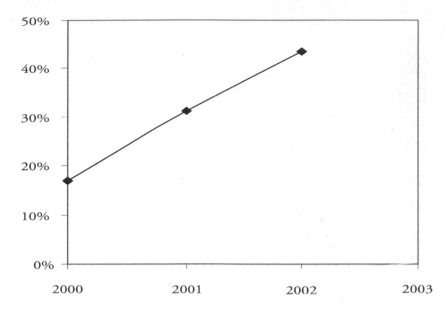

Source: Media Monitors

Figure 4.3 *The rising percentage of print media coverage*
of wind power 'concerning objection'

or negative slant to a renewable project within the article; it merely identifies that the topic of stakeholder objection arose within the article. (This was the case with the coverage analysed in Figure 4.3.)

Second, we should note that the effect of increasing coverage of objection shown in Figure 4.3 may also reflect the natural evolution of coverage. For example, a new and fairly conflict-free wind farm development may only be newsworthy once or twice, whereas a conflict will have much greater longevity as news.

Third and finally, many of the articles about renewable development will be from small, local media outlets focused on local developments, rather than metropolitan media. Thus there will be detailed coverage of planning proceedings, including 'balanced' representation of proponents and any objectors.

It is quite interesting to see that the dominant stakeholder issue area in Figure 4.4, landscape, was of far greater concern than actual economic, physical or environmental impacts. Also interesting is that the most direct impact via property values carries only 10 per cent of that interest. For local residents the two impacts are closely related, although this connection is not apparent from the numbers in Figure 4.4. It is possible that stakeholders are uncomfortable making objections in the social arena based on issues of personal self-interest, such as a drop in property value, especially when the advantages of sustainable energy are known to the wider community.

Interestingly, the prominence of issues like noise, birds and fauna – issues which the wind industry would largely consider to have been resolved – indicates a communication gap between industry and local communities.

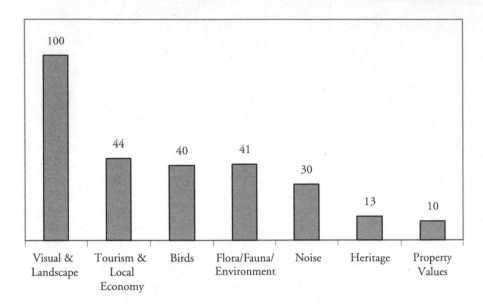

Source: Mallon (2002a)

Figure 4.4 *Seven major stakeholder issue areas identified for wind power: number of references in Australian print media from 2000 to 2002*

Looking for self-identifying stakeholders

Many stakeholders will put up their hand to proclaim themselves given the chance. The media method described above catches some of these but there are also other ways in which stakeholders can be found. Two important avenues are decision-forming processes and decision-influencing processes.

Decision-forming processes

Decision-forming processes include renewable policy discussions or consultations, from planning applications for renewable facilities through to government reviews of legislation. Those putting resources aside to participate are likely to see themselves as stakeholders or potential stakeholders. For example, the Australian Mandatory Renewable Energy Target review received 160 technical submissions and more than 5000 non-technical submissions. Such exercises provide a 'who's who' of supporters and opponents of renewable energy. Less politicized stakeholders such as conservation groups or transport agencies will identify themselves and their concerns through planning applications or when regional planning guidelines are being developed.

Decision-influencing processes

Decision-influencing processes may be generated though pro-renewable energy campaigns which reach out to engage potential stakeholder organizations. That does not mean that one must run a high-profile campaign to discover the stakeholders. However, some of us have done this before and the resultant list of stakeholders may come in handy for this chapter!

Who represents absent stakeholders?

Environmental, social and economic stakeholders can include the very local to those on the other side of the world. The biosphere is impacted by climate change but who is its guardian? People suffering from floods, heat waves or drought are stakeholders, but how are they represented? In order to have complete representation it is important that there be a way to represent absent stakeholders. In practice this falls to: (a) inter-governmental agreements which are translated into domestic law; and (b) the advocacy of non-governmental organizations.

If global and national environmental issues are being addressed by international and national conventions and legislation, we can include the appropriate government agencies responsible for compliance as suitable government stakeholders.

However, the role of government agencies as stakeholders goes beyond that of policing legislation; it is also one of strategic overview and inter-agency communication. If, for instance, repeated rejection of planning for renewable energy projects were to occur – as has happened in the UK – the government's national targets for renewable energy might remain undelivered, requiring the various agencies to review their policies and interactions.

If the absence of some legislation results in interests being overlooked, the breach will often be filled by NGOs. Thus international environmental groups like Greenpeace, Friends of the Earth and the World Wide Fund for Nature (WWF) and their national equivalents may be suitable stakeholders to represent the interests of the wider environment, biodiversity and absent communities. Clearly the influence of such stakeholders is moral and ethical and unfortunately far less binding than specific legislation.

Bringing the results together

No single method will discover all stakeholders. Some groups – like farmers – will not realize they are stakeholders until the industry is up and running, and will therefore remain silent throughout consultations. Others, who may in reality not be impacted at all, could decide that they might be, and become a major voice of opposition.

The matrix in Table 4.2 could serve as a good first cut at the key stakeholders in the renewable energy development debate in many countries.

Table 4.2 A 'type and impact' matrix of renewable energy stakeholders

	Economic direct impacts	Economic indirect impacts	Environmental impacts	Social impacts
Citizen	Employees in industry or supplier Project contractors Site holders	Energy consumers Populations local to projects	People affected by climate change Coastal communities Populations sensitive to extreme weather People affected by air quality, radiotoxic emissions, etc. People concerned about renewable energy impacts on flora and fauna.	People concerned with social impacts such as landscape, transport
Company	Renewable energy manufacturers Supplier industries Project developers Fuel suppliers Financiers Other[1]	Energy consumers Energy retailers Competing energy suppliers Competing GHG mitigation suppliers Groups sensitive to GHG financial impacts: insurers, farming, tourism, forestry	National significance: Farmers	Local residents or visitors
NGO	Industry associations representing any of above Tourist organizations Regional business/development organizations	Associations representing any of the above Unions	International environment groups National environmental groups Ecology and conservation groups National specialist organizations (birds, flora, fauna)	Landscape organizations Hiking groups State or national heritage societies
Government	Local councils Industry development agencies Regional development agencies State and regional tourism bodies Departments of infrastructure, etc. Local authorities Academic centres	Ministries and agencies responsible for: economics, employment, regional development Academic centres	Environment ministries and agencies Sustainability agencies Local authorities Kyoto Protocol (or other) compliance agencies Green Party	Planning authorities Heritage, cultural and conservation agencies Local authority

Understanding Stakeholder Resources

I recently went to meet members of a national organization for bird lovers, or twitchers, as they are known in the trade. With me went an individual from an international environment group; we were there to talk about climate change, renewable energy and wind power in particular.

On the subject of climate change, our host simply pointed to the white board. Apparently a previous meeting in the same room had just discussed changing migration patterns of various species, changes the members ascribed to global warming. Of course they supported action on climate change and renewable energy, we were informed. 'So what about wind power?' I asked. The host reeled off a list of wind projects that local members had asked him to comment on or assess. 'Look, we know the impacts from wind on birds are low, but when we are asked to look at individual project after project, we simply can't cope. We don't have the resources.'

Similar organizations might express the same sentiment on many issues. Groups funded through membership and donations have few staff. They do not have large amounts of money to devote to problems and they do not have a golden carrot waiting for them at the end of a project or process as a project developer does. However, the staff they do have will be very expert in their field and highly committed to their issue.

So we must reconsider the assumption that stakeholders will generally have the resources to adequately assess the real or apparent impacts in their area of interest. At a project level, guidelines often suggest that key groups be contacted and consulted. This implies that the special interest groups have the resources to keep contributing to consultation on an increasing number of projects. We must assess whether or not such assumptions are valid before we ask stakeholders to participate when they lack resources. If not, we can instead find appropriate means of engagement that match their resources.

In the example of the bird-lover group, the stakeholders were seeking a long-term, strategic solution to the bird–wind issues. This led to their involvement in a government-funded project to develop bird and bat impact assessment protocols for wind farms.

An overview of stakeholder resources

We have already set out a matrix of the main stakeholders. In order to appreciate their behaviour and the strategies they may use to address their concerns, we need to build a resource profile. This must be done stakeholder by stakeholder, a process I have started at the end of the chapter. To begin, Table 4.3, which is indicative only, illustrates some general resource characteristics covering specialist knowledge, human resources and financial means.

Table 4.3 *A sample matrix of indicative resources*

	Economic direct impacts	Economic indirect impacts	Environmental impacts	Social impacts
Citizens	SK – Good HR – Good $ – Poor	SK – Poor HR – Poor $ – Poor	SK – Poor HR – Good $ – Poor	SK – Good HR – Good $ – Poor
Companies	SK – Good HR – Good $ – Good	SK – Variable HR – Good $ – Good	SK – Poor HR – Good $ – Good	SK – Poor HR – Good $ – Good
NGOs	SK – Good HR – Poor $ – Good	SK – Good HR – Good $ – Good	SK – Good HR – Poor $ – Variable	SK – Good HR – Poor $ – Variable
Governments	SK – Good HR – Good $ – Good	SK – Good HR – Good $ – Good	SK – Good HR – Good $ – Good	SK – Good HR – Good $ – Good

Note: SK = Specialist Knowledge, HR = Human resources, $ = Financial resources

Some general observations

Individual people and small businesses, especially those local to projects, will have limited specialist skills or knowledge to undertake risk assessment in the likely areas of interest. At a local level, community stakeholders may have only limited financial resources – coming mainly from their own pockets – and limited expertise in specialists areas. Despite this, they are likely to have large numbers of people to call upon to assist in an unskilled capacity.

Larger businesses and companies are likely to have considerable resources; however, they will have less discretion over their use. There is an expectation that these resources must deliver some commercial advantage. They will tend to have expertise in certain specialist areas, but will also have resources to outsource these missing skills if necessary.

The NGO stakeholders may have very strong expertise in their specialist areas. Big environmental NGOs for instance might have one expert in a given field for a whole country. They may be available for a one-off request or to consult on a crucial precedent-setting issue, but are unlikely to be available on an ongoing basis for repeated consultations, project after project, or to follow several big issues simultaneously. Unlike the local groups, they will have discretionary funds to obtain an expert opinion if required. However, these are unlikely to be justified for ongoing consultative processes with a given industry, but would instead be earmarked for strategically significant projects. Industry NGOs will often be better financed than environmental NGOs – but not always.

Governments often appear to have ample resources in all issue areas. The one caveat to this is that governments are also subject to the whims of political pressure. Their decisions are not always rational, regardless of the quality of the advice.

As a final general note, it may be useful to consider the differences between stakeholders that are local and those that are non-local: a local bird-watching society or a national bird association, a local government body or a national agency, a local green group and a national environment group and so on. Local and non-local bodies will have very different resource and expertise profiles.

How might resources influence how a stakeholder acts?

Let us examine how the local citizen might respond to a renewable energy development. As Table 4.3 indicates, local-level community stakeholders are likely to have only limited financial resources and limited expertise in specialist areas such as birds, flora and fauna, heritage assessment or property valuation. However, they are likely to have many people to call upon to assist in an unskilled capacity, and they usually have a formal say in what happens in their local area. Given these assets, they appear to have three options to address concerns they may have.

1   They can accept the assessments undertaken by the developer in the environmental impact assessment (EIA) process, and trust that other specialist government and non-government stakeholders will act if significant risk exists.
2   They can call in bigger specialist NGOs if there are particular concerns, for example wildlife impacts.
3   They can apply the precautionary principle and use the means at their disposal to halt the process until such time as resources or expertise are available for the required assessments.

Most local people may decide to take the first option which means they do not have to do anything. Some might take a step in the direction of option two; but, as we have seen, they may be disappointed with the response, as NGOs have few spare resources to engage at the project level. Which leads to the third option: minimizing risk by minimizing change. This is easy to do since local people usually have access to the planning process. They can gather numbers of individuals to create an impact on the decision. This portends a poor outcome for the renewable energy project developer.

Walking through this simplified example shows how the lack of appropriate options to assess and minimize risk can push stakeholders to block projects. It can also push some community organizations to block renewable energy legislation.

Thus, it may be in developers' interest to look at ways of making resources available to local stakeholders to assist them in gaining independent advice. This might make stakeholders less likely to pursue a precautionary approach. Alternatively, agreement on independent specialist consultants or organizations, which are acceptable to both local people and developers, could avert duplication for EIAs.

A Rationale for Action: Risk Management

One way to conceptualize the developer and stakeholder relationship is by assessing the nature of risks faced by each of these parties.

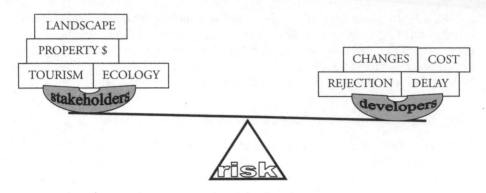

Figure 4.5 *This sample of risks facing stakeholders and developers reveals little overlap*

In Figure 4.5 we can see that stakeholders' risks are concerned with the impact in a given issue area. The main developer risks focus around the project being stalled, delayed or requiring late changes due to one or more of the issues affecting planning consent. Any of these will result in the developer incurring additional costs – a financial risk.

Unlike the developer, stakeholders perceive no reward accompanying the risk. For instance, why would homeowners risk their property value for no good (personal) reason? If compelled, they may protect their interests via various available avenues of objection: 'Great idea, wrong location,' and so on. It is also important to recognize that the onus of risk rightly rests with the developer/owner as the party with the most direct potential gain.

Two solutions appear relevant. First, the risks to stakeholders are quantified, then removed or minimized (at least cost). Second, the benefits to stakeholders are maximized (at least cost).

The second solution is recognized as a key element of German and Danish success with wind implementation, discussed previously in the context of balancing costs and benefits. The first solution implies that there are ways to trade off stakeholder/developer risks; the more stakeholders' risks are minimized, the less the risk of delay or rejection. The important question is how to do this without incurring excessive cost. The earlier in the project cycle that stakeholder risk assessment or mitigation is incorporated, the lower the cost of changes or adaptation.

Effective Engagement and Positive Outcomes

If there is a magic bullet that defines the perfect approach to engagement with all stakeholders it is to treat them all individually and understand their resources and concerns.

At a local level, solutions may often include taking the engagement downstream. It may be that a certain study satisfies the planning authority, but does it satisfy the local people. If not, then what would? What do they need to know for their concerns to be put at ease? For example, could the wind developer commission a wildlife expert trusted by both industry and community, instead of or as well as the one the developer usually employs?

For non-local stakeholders it may be more resource efficient to look at ways to shift the engagement away from a project-by-project basis upstream toward a strategic level. That is to reach multi-project, regional or national agreements on guidelines, practices, experts and so on. Expanding on our wildlife expert example, we might choose an individual acceptable to all major stakeholders, who will define the methodologies for flora and fauna monitoring to be undertaken. Or we could agree on the terms of reference for transport surveys for biomass processing plants. In this way the specialist resources available from some stakeholders may be used more efficiently, while human and time resources for interventions at the project-by-project level are reduced for both developer and stakeholder.

Engaging stakeholders at a strategic level

In both planning systems and guidelines provided by industry, the assumption often exists that engagement starts at the project level. Yet, as we have seen, that level may not suit the resources of many stakeholders. Perhaps the best level to start is with the entire vision, as illustrated in Box 4.2.

Box 4.2 *Strategic stakeholder engagement:*
The Danish pre-planning systems

Krohn (2002) quite effectively explains what may be required for a high-level vision of renewable projects based on experience with Danish policy in this area:

> With a highly visible technology such as wind turbines, the development of models for dealing with public planning (zoning) issues has been very important for many countries' acceptance of the technology. In Denmark the public planning procedures were initially developed though local trial and error. In 1992 more systematic planning procedures were developed at the national level, with directives for local planners. In addition, an executive order from the Minister of Environment and Energy ordered municipalities to find suitable sites for wind turbine siting throughout the country. This 'Prior Planning' with public hearings in advance of any actual applications for siting of turbines helped the public acceptance of subsequent siting of wind turbines considerably. A similar planning model has since been introduced in Germany with considerable success. Other countries are studying these experiences with a view to overhauling their planning procedures.

The Danish consultations on offshore wind power also provide useful examples of engaging with stakeholders at a strategic stage, well before actual projects are implemented. In these cases, assessment of the various use functions of the waters surrounding the concerned areas were conducted including: fisheries, oil and gas fields, military uses and so on. At the same time, stakeholders from the various industries that might be affected, including environmental and ecological groups and tourist industries, were invited to contribute to the assessments. After quite lengthy and in some cases vigorous debates on the issues and sites considered, various areas were identified that would be made available to large-scale (several hundred megawatt) offshore wind developments. Thus the stakeholders were engaged at a strategic (pre-project) level in order to reach agreement on suitable sites rather than only being introduced at the project-by-project stage. Krohn (2002) sums up this experience thus:

> *Around 1997 another set of planning regulations were developed for offshore wind farms, with a central, national authority – the Danish Energy Agency – being responsible to hear all the interested parties, public and private. This 'one stop shopping' method has facilitated the planning process considerably, and is widely studied around the globe.*

Moving stakeholder engagement upstream

The efficiency of the 'one-stop shop' approach to consultation prior to any development, as set out by Krohn in Box 4.2, clearly lends itself to a government-led process. There are, however, options for industry and even developers to move consultation with certain stakeholders at much earlier stages, even without government leadership.

Some expert stakeholder groups may enter the debate only at the planning stage, leading to an engagement via the planning process, as illustrated in Figure 4.6. This obviously provides very inefficient dialogue for all concerned, as by this point the project is very evolved and change will be expensive, and also because the planning cycle is slow and cumbersome. Industry guidelines often counsel the avoidance of such oversights and list some of the stakeholders that could slip though the net of local consultation.

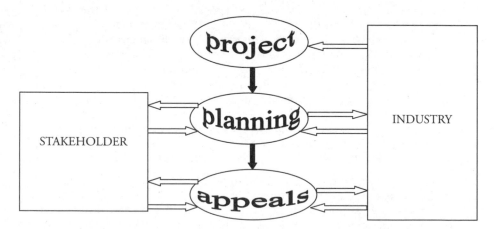

Figure 4.6 *Poorly consulted, overlooked or dissatisfied stakeholders engage with the developer via a formal planning process*

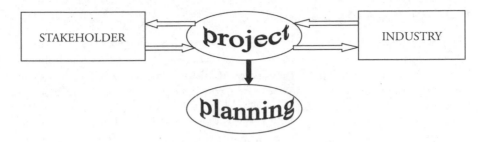

Figure 4.7 *Current best practice at a project level suggests consultation as the project is being developed*

Figure 4.7 illustrates where stakeholder engagement sits as suggested by typical industry guidelines. This is perhaps the only way to engage with most local residents, homeowners and landowners in the absence of a government-led zoning process.

However, many stakeholders could be engaged more efficiently. As discussed earlier, many specialist stakeholders' resource constraints render them unable to take part in project-by-project consultations. Instead they may prefer involvement at a strategic level with the industry or prefer that a large developer agree suitable processes, practices and assessment criteria for multiple projects. Thus, at a project level, their involvement may only be to check that a larger agreement has been adhered to by the developer. If the best specialist stakeholders are not engaged at this strategic level, developers may well find themselves dealing with more locally focused or possibly less expert groups or self-appointed experts – a situation unlikely to benefit the developer or the issue at stake.

Figure 4.8 illustrates a model that may be more appropriate for engaging with regional or national stakeholder organizations.

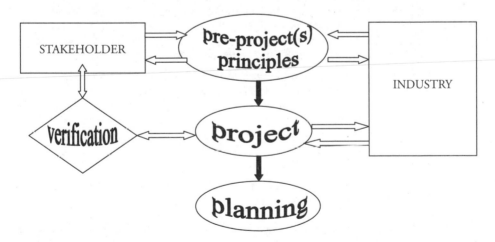

Figure 4.8 *Some stakeholders can be (and may need to be) moved up to a pre-project, multi-project or strategic level to derive in-principle agreements for future projects*

Risk and trust at the grassroots

Most local stakeholders will become engaged around actual projects and will be largely uninterested in the politics of energy. If there is a process designed to consider their interests it will more than likely entail some sort of EIA. Although this may be adequate for the planners and decision-makers, it may be insufficiently rigorous or independent to satisfy stakeholders.

For instance, if an ecology NGO were to accept at face value an EIA assessment from a commercial consulting company hired by the project developer, some might question whether that NGO was executing its responsibilities. Similarly, it might be difficult to expect a landowner to trust a developer-paid consultant's view of possible effects on land values.

This is not an uncommon problem. Although it is a completely different industry, Box 4.3 provides an interesting examination of the lack of trust shown toward an EIA performed for a large planned petroleum development.

Box 4.3 *Oil-shale stakeholders' reaction to an EIA*

In the Gladstone Region of Queensland, Australia, where an oil-shale processing plant was planned for expansion, residential stakeholders were concerned about effects on their house prices. The draft environmental impact statement drawn up by the consultancy firm ACIL for the Stuart Oil Shale Project – Stage 2 stated:

> *Analysis of sale prices of land in the nine parishes over the past 10 years showed that, to date, there has not been any downward trend in rural land prices in the area. In fact, during the past 10 year period, fruit farms have fared well with an upward trend in the value per hectare in keeping with regional trends.*

However, local residents and farmers were concerned this was not the case. The Yarwun Targinnie Fruit and Vegetable Growers Association Inc. put together funding to commission a strategic plan for the area which included an assessment of impacts from the oil-shale development on property values. The study by ImpaxSIA revealed that of 190 rural and town lots:

> *At the present time there are approximately 30 properties openly listed for sale and numerous others which have been withdrawn from sale following a total lack of interest from potential purchasers... One prominent agency reported 17 properties listed for sale in the Targinnie area with no potential contracts likely to be signed. Some rural land agents reported it was a waste of time and petrol money to drive out to Targinnie to erect a 'For Sale' sign.*

It is interesting to note that the oil-shale project developers Southern Pacific Petroleum and Central Pacific Minerals contributed $75,000 towards the association's study. One assumes that the developers were comfortable with the independence and integrity of the ImpaxSIA consultants. Time and money may have been saved if an agreement on a mutually trusted assessment group and/or remit had been made at the outset.

Source: Mallon (2002b)

Regardless of the point at which stakeholders are engaged, it is essential they accept an EIA's integrity. If they do not, they will either contest the EIA by introducing their own assessment, block the process with arguments based on the EIA's validity or attempt to challenge the process through other means or issues.

The most obvious route to stakeholder acceptance is to consult and reach agreement with stakeholders on the terms of the EIA and the organization which will undertake it.

A Closer Look at Some Key Participants

There is only so much you can put into a table! The matrix discussed earlier is an important way to catalogue stakeholders and ensure we know where to expect stakeholders to emerge even if they have not popped up yet. However, it is also worth taking a more specific look at some significant stakeholders, and making some general recommendations for action.

The citizen

Individual citizens may be the smallest unit of society but they can also be among the most influential – such as when they vote or write a letter. Their source of engagement with renewables will generally be from a project level impact or from a point of principle (such as the environment).

Working optimally with project-impacted individuals relies on the art of community consultation – either by government or industry – and should be undertaken as thoroughly and expertly as possible. People who become involved because of a principle will have to be engaged by industry, NGOs or government and given avenues through which to have their voice heard (see Figure 4.9).

In terms of taking action to improve chances for positive outcomes, it is important to be on the front foot with citizens and be the first to make the approach.

Individual investors

Individuals may invest in renewable energy companies and therefore be financially engaged with the renewable energy industry. This tends to be quite a limited group since few investment opportunities in renewable-only companies exist, either globally or locally. People making the extra effort to find and invest in renewables are likely to be highly principled and motivated. They are also likely to be reasonably well educated and empowered and so would make excellent allies. However, they are hard to locate.

These people may be found through the companies that they are investing in, if those companies have ongoing contact. For more high-profile and upbeat companies this can also provide an excuse for a positive and dynamic engagement with their investors to elevate them beyond silent partners.

Renewable energy from wind, solar
and biomass avoids the greatest
source of greenhouse gas—
power stations burning coal.

Join the campaign for 10%
new renewable energy in 2010.
The future depends on you!

Source: Created for the Australian public during the MRET (Mandatory Renewable Energy Target) policy
review period

Figure 4.9 *A public outreach website used to engage individuals
and encourage action*

Renewable technology manufacturers

These companies tend to support the highest renewable targets they consider
politically feasible. This support can be undermined if some of the bigger companies have mixed interests in areas that may be adversely affected by increases in
renewable energy production.

As we have seen in previous chapters, renewable industries need to be aiming
for exponential growth. However, high targets can be undermined if some of the
smaller companies have quite modest visions for their sector's future. A doubling
of demand for renewables spells huge success for a small domestic manufacturer.
However, given that such industries are always starting from a small base, meaningful environmental benefit is unlikely from such a gain. Nor will it put the
industry on a footing to create large-scale manufacture and compete internationally. Thus there is a risk that such industry stakeholders could make poor negotia-
tors for the industry as a whole.

Many of the larger companies are international and likely have top-down
codes of practice that limit freedom of action to lobby or campaign overtly in a
given country. This may be because of shareholder sensitivities or disclosure rules
in their home countries. Alternatively, some transnationals may have positioned
themselves as taking a strong stand on renewable energy issues that can be activated in the new market.

Such companies will generally shy away from being overtly critical or vocal
about governments to avoid prejudicing future government assistance for manufacturing plants and the like. However, the larger companies do have the ability to

invest heavily to open up new markets and this can include helping their representative industry organizations who can do the lobbying and campaign work instead.

Although in principle these companies will be key proponents of renewable development, there is a need to take action to bring them together to ensure that they speak and negotiate with a unified voice. *Rule Number 1: Don't try to promote your own renewable by criticizing another renewable.*

Renewable fuel suppliers

Groups such as farmers or agricultural produce companies that control wind and biomass resources are generally absent in early policy debates – indicating that this group does not self-identify and is hard to mobilize. In fact, when establishing new markets, most of these resource holders may not be aware of the resource they have or of its value. For example, once somebody buys a solar hot water heater they effectively become a renewable energy harvester – they are part of the renewable industry! There may be very large numbers of such stakeholders available, who can be activated with suitable strategies.

Once established and experiencing significant real or potential income/benefits from renewables, the asset-holding stakeholders can become quite powerful. Although landowners and growers often have quite traditional and conservative political allegiances, as business owners they are very economically attuned and will treat any real economic opportunity seriously.

To make this group into effective stakeholders that represent their own interests well, others may have to take action to identify, educate and facilitate these stakeholders' engagement in the policy forming processes. Project developers will often have access to such stakeholders and potential stakeholders. They may be consulted through appropriate associations.

Renewable developers

There tend to be two types of developer: single-interest and multiple-interest. We will concern ourselves with single-interest developers who are only or primarily focused on renewable energy development. As for multiple-interest developers, we will be concerned with those that have diverse interests stretching beyond renewables. For example, in many countries the existing fossil fuel or nuclear companies are expanding into renewables.

While single-interest developers are likely to unequivocally support high renewable development targets, multiple-interest developers may have internal conflicts of interest that mitigate against large increases in renewable energy markets. State-owned companies will often have to toe the party line and adopt a position consistent with their government owners.

One other sensitivity here is market control. Some companies may actually support a slowdown in renewable sector growth to preserve a market advantage or to avoid a period of market instability that threatens their position. For example, in the case of Australia's MRET review we saw diverse posturing among renewable developers with positions advocating target levels ranging from 2 per cent to 10 per cent.

It is important to take the necessary action to ensure a consistency of voice on renewables if possible, to ensure the industry is seen as a credible whole. Given the mixed-interest issue, it is essential that renewable developers do not leave it to the largest companies to do their lobbying as they may represent other interests besides renewables.

Finance companies

Many big financial houses are now engaged in the renewable energy industry, including big names from Europe, North America and Asia. Many transnational finance companies are now familiar with and active in financing renewable energy projects and are seeking new markets.

Finance companies are likely to be early movers, perhaps ahead of domestic players, if renewable energy development policies are put in place. They have significant influence in big business arenas and therefore government circles. Although their representatives may attend lots of conferences it may be harder to get them to be active lobbyists on renewable energy issues until markets are actually underway.

These are influential industry participants. Even if they are not yet active in a particular country the mention of their name in connection with the international renewable industry can surprise and enlighten decision-makers.

Institutional investors

Institutional investors can vary widely from large and silent monoliths to responsible ethical investment groups, and on to potentially quite militant union-based investors. The decisions that such investment houses make send strong signals that are useful if they are investing in renewables.

The power of the institutional investors lies in their wealth and placement of that wealth. Their engagement with government is infrequent. For example, there were no institutional investors noted among Australia's MRET review submissions, even though many may have sustainability criteria in their lending policies.

Nonetheless institutional investors will be experienced in political and policy processes and can provide useful counsel to other players in the renewable energy industry. Institutional investors are increasingly being targeted as ethical arbiters by environmental groups because they are a powerful group that is answerable to individuals.

Although their influence may not be as direct and tangible as in the company arena, institutional investors that are encouraged to participate may be a discrete voice which can gain the ear of government and parties of influence. They also have the ability to directly communicate to their members on issues they deem relevant. In terms of action, they can provide valuable experience and insight into the wider industry.

Competitor fuel and power suppliers

Any market has an array of competitors. It is probably worth tackling a few of the major ones in order of importance.

Coal industries

The coal industry is massive, established, wealthy and also highly polluting. Despite the introduction of oxymorons such as 'clean coal', the bottom line is that coal is carbon and when you burn it you get carbon dioxide. There are always new ways to get slight increases in efficiency, but coal cannot escape being the world's big greenhouse problem.

Coal industries have good cause to feel threatened by other fuel sources in both established and new markets. In established coal markets such as the European domestic market, coal has been undercut in price first by gas and then by wind. German coal industries are propped up by ongoing subsidies that even the European Commission cannot seem to bring to an end. The coal sector in other countries will not want to let the same happen to them.

As for new coal markets, the international power supply industry is focused on industrializing countries around the world that will double the current installed power capacity of the planet in just two decades. This growth represents the life blood for many industries like coal and nuclear power, and huge resources are being spent trying to lock these markets in. The fact that this represents a climate time bomb may be secondary in their thinking.

Thus the global coal industry seems to be faced with an apparently irreconcilable problem. The world wants to de-carbonize and coal is carbon. The US and Australia, two big coal producers, are both focusing on geosequestration, which involves the end-of-pipe capture and long-term storage of carbon dioxide. In principle there ought to be little threat to renewables since the costs of geosequestration are extremely high and the places to sequester CO_2 quite limited. However, one should never underestimate the lengths people and industries will go to when faced with issues of survival.[2] Thus renewable energy proponents may not find themselves engaged in a clean fight with industries such as the coal sector.

Gas and oil

The gas industry is a relative newcomer to the power sector and, in a similar fashion to renewables, its advancement often requires policy reform and the removal of access barriers. The European example shows how the growth of gas and renewables are complementary in both emissions reductions and power management. Gas turbines can increase and decrease output quickly making them ideal for covering fluctuations in variable renewable sources.

In some regions gas has many advantages over renewables, although cost and availability may not always play in its favour. However, as alluded to in my previously mentioned joke about the two people running away from the leopard (I don't have to run faster than the leopard, I just have to run faster than you!), the gas industry may see renewables as a competitor, not an ally, and act accordingly. Gas may be competing for market access against renewables and may see its fortunes to be improved if renewables are constrained. It is also an unfortunate consequence of some electricity markets that if they are based on short-run marginal cost (i.e. the cost of fuel not capital) then new renewable generation can end up displacing gas instead of displacing coal or nuclear power. This is clearly a failure of the market structure and it must be seen as an opportunity for renewable and gas industries to work together to correct the situation.

In the Australian gas industry, the decision was taken to lobby against renewable industry development policies in favour of emissions trading, a policy the gas industry considered to be more cost effective in the short term. Similar problems have been encountered between bio-fuels and the oil supply industry where one can see that displacement of oil-derived products is an immediate threat to oil suppliers.

Large hydro

Large hydro installations can provide power on demand, which is useful for high-value peaking load and therefore already has quite a strong economic value. As we have discussed earlier, large hydro needs to be distinguished from new renewables for two reasons. First, it has been around for long enough to be considered conventional energy and second, it cannot be taken for granted that it is environmentally sustainable.

In many countries, the development of new large-scale hydro is becoming increasingly difficult due to environmental and social impacts. It is likely that any future projects will have already been identified and will not be affected by renewable energy legislation. However, it is also likely that the large hydro producers will nonetheless try to have renewable energy legislation include their existing or planned projects as this provides a bonus revenue stream. The scale of large hydro projects – built or planned – means that, if they are included in renewable energy policy measures, they will tend to dominate the schemes and financial flows for several years.

Nuclear

The future of the nuclear industry hinges very heavily on new projects. The lack of projects in its homelands of Europe, Eastern Europe and North America means this industry is fighting for its survival. The nuclear industry has been linked in the British Press to the UK anti-wind group Country Guardian via its co-founder Bernard Ingham who has been involved with the nuclear industry. And I myself have witnessed the industry being highly critical of renewable technologies in international fora such as the United Nations Framework Convention on Climate Change (UNFCCC). It is a highly resourced industry and should not be dismissed.

As nuclear generation is more expensive than conventional energy and some renewable sources, it must call upon other advantages to make its case. In nuclear countries the industry is vaunting itself as a climate change solution. Although nuclear energy proponents believe it to be technically proven, it nonetheless has many detractors. Nuclear plants produce reliable energy, but long-term waste storage is not considered to have been solved and indeed there may not be a solution. A more immediate problem is the build-up of weapons-suitable plutonium in Asia and other new markets as a direct result of running nuclear power stations.

Conventional energy generators

In principle generators would be expected to be neutral with respect to fuel source. For example, if burning bio-fuels provided competitive advantages these stakeholders would likely have little allegiance to fossil fuel suppliers. It is true that generators with diverse fuel sources are more likely to embrace renewable sources and engage with (moderate) pro-renewable policy-making. However, generators with a single fuel source must be convinced of the opportunities or they will simply see renewables as a threat. The depth of their links with pre-existing fuel suppliers may make this transition difficult.

Energy transmitters/grid management

If the entities that manage the grid are separate from energy sources, then, in principle, they ought to be neutral with respect to sources of power put into the grid. However, in practice, they will often be unimpressed by intermittent supply sources and can place onerous barriers in the way of renewables.

When acting to deal with this challenge, early constructive engagement is key. Much has been done already in pioneering countries and this can help inform grid operators on how to best manage renewable energy.

Energy retailers

Energy retailers are the final step in the supply chain that brings energy to the end consumer. Energy retailers will react very differently to renewables depending on whether they are in a competitive environment or not and also whether they are vertically integrated or not.

In markets where they do not have to compete for customers, retailers will be less sensitive to new generation even if it causes their prices to increase. After all, their customers cannot leave them.

If energy retailers are in competition, their chief concern is whether renewable energy policies would push them into a competitive disadvantage. This happened in Germany in the late 1990s when retailers were put in a position of competitive disadvantage until the cost burden was re-distributed by policy changes.

A market which allows vertical integration of retailers may create some pressures for companies to protect their upstream conventional generation assets and prevent the market from being opened up to third-party renewable energy generators. In this case they may act along similar lines to energy generators as discussed above.

On the other hand, these stakeholders are increasingly positioning themselves as energy companies, signifying less and less dedication to a particular fuel source or energy types. This can mean that energy companies take a more neutral approach to energy sources. In markets with contestable customers, some may try to position themselves as cleaner or more focused on renewables. However, they will be sensitive to price, as price increases mean fewer sales.

Those seeking to establish new markets for renewables will, of course, be challenged by strong links between retailers and existing generators. These retailers may be lobbied to follow similar positions along the lines of their conventional energy suppliers (discussed below).

General energy users

In all countries from G8 (the group of 8 leading industrial countries) to G77 (the broad alliance of mainly developing countries), the link between energy use and economic development has appeared to be unbreakable. Thus, most governments will work very hard to facilitate low prices for commercial energy and fuels. Energy is often provided at low cost for most residential consumers and industry sectors. Energy security and reliability are generally more significant concerns than price.

Small changes in cost due to the inclusion of renewable generation are unlikely to put off consumers that are not exceptionally high users. This result can be promoted through polling or garnering support from industry groups or high-profile companies.[3]

Energy-intensive users

Companies which smelt metals such as aluminium are very sensitive to power costs. Heavy manufacturers such as car makers are also big power consumers. These industries often have special commercial or governmental arrangements for energy provision, which permit them to pay lower rates or be excused from some of the levies normal consumers pay.

Companies for which electricity is a considerable expense are often opposed to anything that might affect the price of energy. Although in principle they will oppose any cost increase, their underlying sensitivity is about competitive disadvantage. Thus the context in which companies are operating will be important.

In large markets like the EU or US, where most of the product is being consumed internally, there may be an argument for a policy structure that will affect all competitors equally, or shield these businesses from any loss in competitive advantage from imports by the use of commensurate trade tariffs. In some cases a waiver from the cost impost of renewable energy may be the most sensible solution to avoid a head-on battle with an energy-intensive company or industry.

Unions

Obviously, different unions will often have varying positions on renewable energy. However, all unions will be sensitive to employment levels and any risk of employment loss. They may be influenced by any perceived employment risk to their industry that may be conveyed by the major employers. They will also be interested in opportunities for unionizing new industries. Moreover, unions often have progressive environmental policies. However, tensions within a sector can be transferred to a union, for example, tensions between coal and renewables.

In terms of action, it is very important to take the necessary action to address unions' tensions or conflicts regarding impacts of renewables on the energy sector.

Farmers and landowners

Farmers can be incredibly independent-minded and progressive, such as those embracing renewables in countries including Denmark, Germany and Israel. However, they can also be more conservative and cautious and are often not aware that they are one of the primary victim groups of climate change. Furthermore, those with no possibility of having a direct financial interest in renewables may be resentful of other farmers who have.

In terms of taking action on this point, there is value in engaging and building relationships with farmers and landowners at a project level and organizational level if possible. Preferably this is done in parallel with environmental groups that provide information on the climate change impacts on agriculture.

Non-governmental organizations

Renewable energy NGOs

Renewable energy associations can of course be expected to be great supporters of pro-renewable energy policies. However, in new markets they will have a few challenges. They are likely to be small or, worse still, non-existent. In some cases, renewables may find themselves represented by other associations which have mixed agendas that conflict with the promotion of pro-renewable policies. The associations are quite likely to be fighting among themselves – often because they are competing for limited resources.

Both of these points will take a heavy toll on renewable associations' political influence. With no industry to show, no workforce to flex and little influence with voters, the message of these associations can sometimes sound like a list of promises. It is, in effect, a promise of carrots with no accompanying stick.

In the early days of their development, such associations will need to take action to make strategic links with natural allies that already have influence within the society. We will discuss this in detail in the next chapter.

Sustainable energy NGOs

Although it may seem to be merely a semantic distinction, sustainable energy NGOs are distinguished here as addressing the increased use of energy efficiency and gas (which may or may not be limited to co-generated gas) in addition to renewable energy technologies.

Sustainable energy NGOs have other considerations that may compromise their unmitigated support for renewables – or at least mean they have other foci. There may be non-renewable industry members/groups within the sustainable energy NGOs that are more established and wealthy. This can increase the overall influence of the NGO but may also allow these powerful members greater influence over decision-making than renewable energy members.

Thus, it must be understood that sustainable energy NGOs are useful and potentially powerful advocates, but may have internal conflicts of interest that affect their level of support for renewable energy industry and policy reform.

Local business/economic NGOs

Many local communities have organizations, or are branches of larger organizations, that allow local business people or professionals to meet, network and collaborate on assisting the wider community.

Renewable energies can provide a benefit to local communities through installation and ongoing operation of renewable technologies or fuel collection, and where relevant through manufacturing capacity. Unless there are projects already in process, local business NGOs are unlikely to be aware of the community's renewable resource and its economic implications.

In terms of actions, a programme of liaising with local economic NGOs can help to overcome uncertainty about renewable energy and provide important and influential local contacts and allies for wider policy processes.

Fossil fuel NGOs

Industry associations can often take stronger positions than the companies that fund them. The companies have their credibility on the line and perhaps also a brand to protect. An industry NGO has no such constraints and so can be far more vociferous in its campaigns to limit the development of renewable industries.

A fossil fuel industry association by its very name does not allow for diversity. Whereas a member company may see an advantage to diversifying into renewables and keeping an open mind to new opportunities, the fossil fuel NGO will see only the threat to the fossil fuel industry as a whole, and any other interests of their members are beyond its concern.

Fossil fuel NGOs are likely to be extremely influential with government as their products dominate most economies. They will also have significant wealth with which to resource their causes through PR companies, lobbyists and getting influential consultants to write reports. There may be little room to persuade such groups to be on side with renewables or take a neutral stance.

However, these groups are also answerable to their membership. Members (which do have brands and are more visible in the community) can call the associations to take more progressive or more neutral stands.

In summary, there is no way to out-gun large conventional energy industries at their own game. However, the minimum that can be done is to take action to carry out a basic level of lobbying to ensure politicians are aware of a different viewpoint, and provide correct fact sheets and information on issues. Moving early can also help set the standard of the debate about clean energy. And finally fossil fuel NGOs can be reined in by their members if it becomes necessary.

Energy generator NGOs

Generator associations may be specific to a fuel source, or their fuel source may be diverse. The former can be expected to resist loss of market share; the latter will potentially be more neutral.

In terms of actions, renewable generators as members or industry colleagues should strive for healthy working relationships to encourage progressive or neutral positions.

Energy retailer NGOs

The complexion of any energy retailer association will reflect the complexion of its members – who in principle need not be attached to any particular fuel source. Renewables may have some representation with retailers if they have some existing renewable interests such as the supply of green power.

Renewable retailers must take the necessary action to gain membership or representation with these NGOs, to ensure such associations do not take a defensive attitude that resists change.

Business NGOs

Chambers of commerce or other broad-based national industry and commerce organizations tend to be very conservative in their positioning and have a big-picture focus. On issues of climate change they will tend to focus on the Kyoto Protocol or carbon trading end of the debate rather than become too technology specific.

In practice, because there will be winners and losers in most areas of climate change action, such groups are often unable to make a stand one way or the other. If they do take a stand against renewables because of the increased cost to business as a whole, for instance, then they can be very influential on government. Exceptions are NGOs that represent environmental businesses, which of course have a proactive position on environmental action and will want to be seen to take positions.

Although in principle they should be neutral, the influence of these NGOs is important. These organizations should be actively informed and engaged by the renewables industry as early as possible.

Environmental NGOs

Environmental NGOs can actually be the most dynamic proponents for the push to get renewable energy policies underway, as Chapter 9 on Spanish energy policy demonstrates. Bigger environmental groups that engage on climate issues will invariably do so in a positive way and will often have pro-renewable energy policies. Note that their agenda is to employ renewables to solve climate and pollution problems, not build the renewables industry for its own sake. Smaller NGOs may be more locally focused on ecological issues which can potentially result in an anti-development reaction to renewables.

The Green vote and environmental credibility are increasingly influential in many countries and these NGOs hold the keys. Furthermore, environmental NGOs usually work through building a public support base which is potentially a great asset for the renewables industry.

Environmental NGOs' understanding and support for renewable energy cannot be taken for granted. It is important to take the necessary action to develop good working relationships with these groups.

Social and conservation NGOs

These NGOs address issues that affect people and their social space and include landscape and cultural heritage NGOs, churches and local community organizations. They are critical stakeholders that may be tangentially affected by climate impacts but may not have formed policies. They may be less adversarial or activist than environmental groups, perhaps taking up a role of advocacy with government.

Conservation NGOs will tend to regard any development as detrimental (as something is changed rather than conserved). However, this does not rule out their acceptance of balanced interests. Active dialogue can help develop their appreciation of the contexts.

In terms of action, early engagement of these critical stakeholders is essential as they will often be called upon for advice by concerned citizens local to proposed projects.

Professional NGOs

Professionals such as doctors, lawyers and engineers have associations that are respected by wider society. Polling indicates that awareness of environmental issues increases with income and education, so professionals will often be some of the most environmentally aware people in society. Some – such as doctors – may have a direct interest in climate change due to the current or anticipated health impacts, and their NGOs may have environmentally active sub-groups. Others may be interested in an industry angle.

Professional NGOs are unlikely to self-start in support of renewable energy, but will usually have a process for addressing requests for support. This can be a worthwhile exercise since professional groups can make powerful public and private allies for a new industry.

In terms of action, these groups can be asked for specific support on key issues, for example, a piece of legislation.

Anti-Renewable NGOs

These come in three broad groups:

1 genuine members of communities that are concerned about specific issues;
2 think tanks or industry-funded organizations that have a specific anti-renewable agenda on energy or environmental issues; and

3   groups with mixed agendas which are less issue- and more cause-focused;
    for example, many anti-wind groups purport to support wind in the right
    locations, but have been seen to criticize wind power on any grounds and in
    almost any location; thus the cause may have become an end in itself as far as
    wind development is concerned.

Unfortunately these three types of groups may coalesce or overlap, making it
difficult to deal with genuine issue-based concerns and equally difficult to demonstrate secondary motives if they are present.

   These things considered, it is important to take action to address community
concerns early and thoroughly. It is also important to correct wrong information
provided by such groups through normal media correction, but also through legal
means if it becomes persistent, deliberate or malicious. It may also be critical to
ensure that supporters of clean energy are heard at a local level.

Government

It is useful to consider two different types of governmental stakeholders: politicians and bureaucrats.

Government – political

This group includes politicians, advisers, political parties and their traditional
allies. These stakeholders will tend to be quite focused on public feeling, marginal
seats, voter sentiment and the attitude of different demographics or traditional
allies. The views of these groups are directly related to their tenure and therefore
their decision-making power. They will be balancing these indicators with the
technical advice from their bureaucracy.

   These groups also tend to have signature issues upon which they focus, such
as job creation or environmental protection, providing a filter through which they
will approach new issues.

   Political decisions may not appear rational at first glance. Thus issues must
be considered from a political angle – rather than technical or economic – for the
decision-making process to be understood.

Government – bureaucracy

The bureaucracy and government agencies are in principle non-partisan in many
countries. Their mandate will be to provide the best outcomes for the constituency in their issue areas, be it industry development, energy security or environment.

   Although politically neutral in principle, every employee will have a fairly
clear idea of what is and is not possible given the likes and dislikes of his political
masters. Bureaucrats may also be receiving less than favourable information on
renewables from other sources. Nonetheless, the skill level or technical expertise
of bureaucrats cannot be taken for granted. It is up to industry lobbyists to ensure
that they are kept abreast of renewable energy developments.

In terms of action, solid networks and good information flow are crucial to ensuring that these bodies are producing sound information and recommendations for decision-makers.

Conclusion

In this chapter we have shown how renewable energy, as a new and expensive product entering a highly competitive market has many potential friends as well as some stiff opposition. Whether they are concerned with a lone citizen armed only with a laptop or the global fossil fuel corporate giant with seemingly unlimited resources, the renewable energy industry must find ways to engage stakeholders effectively to maximize support and minimize opposition.

We saw the various ways that stakeholders can experience impacts – direct economic, indirect economic, environmental and social. We combined these impacts with the various stakeholder types – citizens, companies, NGOs and government to populate a stakeholder matrix. We also found that stakeholders can be identified in different ways: by using first principles (i.e. identifying them by stepping through what would happen if a renewables project were developed); using the media to identify stakeholders; or looking for self-identifying stakeholders during decision-forming processes and decision-influencing processes.

We repeatedly saw the value of identifying and engaging stakeholders before they identify themselves. Early engagement can influence the process, whether it is through providing correct information and reassurance to local citizens, or establishing the standard of debate when anticipating opposition from powerful fossil fuel lobbies.

We also looked more closely at the main classes of stakeholder groups to estimate what their various human, financial and expert resources may be. We saw the value in engaging some stakeholders at higher and more strategic levels of project and policy development to prevent a potentially wasteful project-by-project approach. And we learned how finding mutually agreeable consultants can stave off costs and neutralize suspicion and conflict. We demonstrated how lack of appropriate options can push some stakeholders to block projects or legislation.

The value of minimizing stakeholder risk and maximizing benefits was illustrated, demonstrating that taking early action in the project cycle can reduce costs of change or adaptation. We also saw how countries such as Denmark engage in full stakeholder consultation as early as the high-level planning stage of their overall vision.

Finally we took a close look at a variety of stakeholders and made recommendations on how best to engage them, from farmers and landowners who may be quite unaware of renewables' benefits, to threatened industries such as nuclear or coal which may actively oppose renewables.

Notes

1   There will of course be stakeholders that defy simple cataloguing. Some interesting examples include the military, transport agencies, aviation agencies and television transmitting companies.
2   In Australia, for instance, it is a matter of record that Rio Tinto's chief scientist advised the government that geosequestration would be a tenth of its actual cost and several times lower than its predicted costs.
3   For an example of how to cost an increase in renewable energy supply see *The Cost of Increasing the Mandatory Renewable Energy* (Reardon and Mallon, 2003) a report carried out for AusWEA. www.transitioninstitute.org

References

Beder, S. (1998) 'Public Relations' Role in Manufacturing Artificial Grass Roots Coalitions', *Public Relations Quarterly* vol 43, no 2, pp21–23; also available at www.uow.edu.au/arts/sts/sbeder/PR.html

Country Guardian (2003) 'TV Botanist Opposed to Wind Farms', reproduced from *Western Morning News*, 26 November, on the website of Country Guardian, the UK's National Campaign which opposes many wind turbines, available at www.countryguardian.net/Bellamy.htm

Krohn, S. (2002) *Wind Energy Policy in Denmark: 25 years of Success – What Now?* Danish Wind Industry Association, Copenhagen

Mallon, K. (2002a) 'WindPower the brand', conference paper to the Australian Wind Energy Conference, Adelaide, July

Mallon, K. (2002b) 'Issues relating to stakeholder interaction with wind power development', issue paper for University of New South Wales Energy Policy Group, 17 July

Reardon, J. and Mallon, K. (2003) *The Cost of Increasing the Mandatory Renewable Energy Target*, AusWEA, Melbourne

5

The Politics of Achieving Legislation

Karl Mallon

Once you know what you want, then you can ask for it.
But actually getting it is another matter.

Introduction

Before we embark on this chapter, we must note that we are about to move into a new universe. We are leaving a Newtonian universe, where things are solid and behave in a known and predictable way, for the quantum land of politics where nothing is guaranteed and all is probabilities! We now know with some accuracy which policies work and which do not. We know what we want. But, as we seek to have this implemented, we enter a maelstrom in which our influence is just one force among many and our issue just one in a galaxy of others.

There are no guarantees in politics, but we can endeavour to maximize chances for a positive result and minimize chances of a negative result. Therein is our task.

This chapter is to political theory what chewing gum is to cuisine, but it is a good start for our purposes. This unholy mix of strategies and tactics has been gleaned from rubbing shoulders with strategists and communications professionals while working for renewable policy change in many countries.

The challenge with political campaigning is that one must almost certainly be a participant to glimpse and understand the behind-the-scenes push and pull of stakeholder manoeuvres and political influence. This makes it difficult to describe campaigns that one is not directly involved in. Thus, in this chapter, I will illustrate each point by drawing on my recent experiences in the work with the Australian Wind Energy Association (AusWEA) to have the national Mandatory Renewable Energy Target (MRET) expanded. It is a very useful example indeed, as this is a major renewable industry campaign that ran for more than 18 months and dealt with significant challenges and barriers. However, the basic thesis of this chapter will be valid for campaigns generated from industry, government or NGOs in most countries.

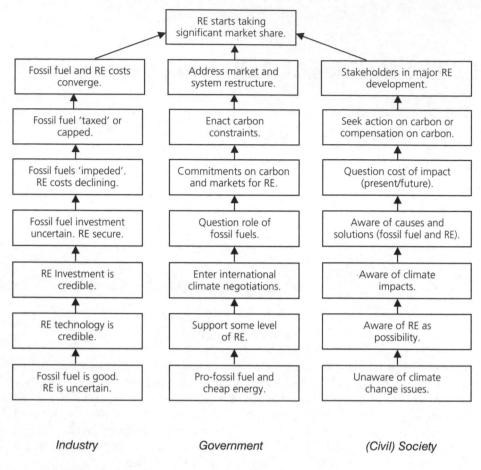

| Industry | Government | (Civil) Society |

Note: RE = renewable energy.

Figure 5.1 *Conceptual diagram showing the paths that industry, government and civil society travel en route to major renewable energy development*

The three pillars of change

In order for renewable energy policy to be legislated it must be accepted by government, industry and society. If any one of these groups is not ready for each step, change will be very hard to achieve. Therefore, it is up to each of these three groups to bring the others along with it. For instance, a government may be keen to embark on indigenous energy production, but the industrial human capacity is minimal and the social understanding is equally weak. In this case we might see a government do the campaigning.

Is It Harder to Shout 'Yes' than 'No'?

As we will see from the Spanish and UK case studies, instances of opposition to renewable energy have arisen. This is despite perhaps three decades of work by many thousands of researchers, engineers, scientists, campaigners and politicians to find sources of energy that can be developed in harmony with the environment. As with most things in life, it is easier to say 'no' than it is to say 'yes'. Similarly it is easier to block and hinder a process than to clear a path and find solutions.

The irony of campaigning for reforms in energy legislation is that the environmental movement finds itself in the novel situation of doing the problem solving. Sometimes this can leave campaigners frozen to the spot and quite unable to rise to the challenge. With the emergence of anti-renewables groups it sometimes appears that some environment groups or parties can find themselves siding with the 'No' camp out of sheer habit.

Nevertheless campaigns for solutions can be successfully accomplished. Campaigns for progress, such as the vote for women or civil rights, are some of the world's most celebrated social campaigns. We are accustomed to seeing development, especially industrial development, as always harmful to the local environment. There is no doubt that renewables affect the local environment, but it is in the interests of a much larger environmental success.

The Spanish case study in this book reminds us that locally oriented groups may be unaware that global risks lead to local impacts. Therefore it becomes essential that the larger groups and agencies – familiar with the climate change problem and its impacts – champion the cause of solutions. They must also provide leadership and guidance for the smaller environmental groups that do not have the same level of expertise or exposure. In solutions campaigning, the key word is leadership.

Solutions campaigns must not flinch from the touchstone of any campaign, which is to inspire strong feelings on an issue. The one thing that will not lead to success is apathy and disinterest. Consider the film made for AusWEA, described in Box 3.6, which is designed to be emotive, not technical.

Politics is like a game of football

Ultimately it is politicians who actually make decisions about new legislation and vote that new legislation through. So finally we must climb out of our ivory tower of concepts and strategies and move out into the cut and thrust of real-life politics.

In the previous sections we noted the types of policies needed to secure a functional renewable energy industry. We also identified how different parts of society and various sectors will be affected by the transition towards renewable energy, and the pathway government and industry must tread in order to accept a significant role for renewable energy in the country.

We might think of winning legislative change as a game of soccer or football. The ball is our package of policies. The goal is getting these to be legislated. However, there is a twist in this game. Each team can have as many players as they can muster and these players can leave the game or join the game as they see fit. And in an even more perplexing twist, some players can change sides if they want!

Box 5.1 *A look at political behind-the-scenes influence*

This is a transcript from an Australian radio programme called 'PM', broadcast on both local and national Australian public radio on 7 September 2004. It shows how leaked documents reveal fossil fuel influence in an important energy White Paper released by the government. The transcript runs as follows:

MARK COLVIN: Remember the row back in June when the Federal Government released its energy White Paper and the renewable energy industry saw it as a huge let-down?

The White Paper recommended a big investment in geosequestration – pumping carbon emissions from power stations and other polluters into holes in the ground.

The people who make windmills and solar panels, among others, got very little, and believed they'd been passed over.

Well, now the ABC's [Australian Broadcasting Corporation's] Investigative Unit has obtained leaked meeting minutes, emails and memos which suggest that behind the scenes, the fossil fuels industry had a huge say in what went into that White Paper.

With both sides now out wooing the environmental vote, the documents have an added political significance at the moment.

Andrew Fowler of the Investigative Unit reports:

ANDREW FOWLER: The group of 12 companies were hand-picked by the government to form LETAG – the Lower Emissions Technical Advisory Group.

Fossil fuel producers – Exxon Mobil, Rio Tinto, BHP Billiton and high-level fossil fuel users and generators – Alcoa, Holden, Boral, Amcor, Energex, Edison Mission and Origin Energy were part of the government's exclusive invitation-only group.

Clive Hamilton is the Executive Director of the Australian Institute and a critic of the coalition's energy policies.

CLIVE HAMILTON: This leaked document provides a remarkable insight into how the policy agenda is really set under the Howard government. It's quite clear now that when the Prime Minister wanted new policy directions to deal with climate change, he decided to call a secret meeting with Australia's biggest polluters and said 'Tell me what I need to do.'

ANDREW FOWLER: The energy White Paper – 'Securing Australia's Energy Future' – was released this year and detractors immediately accused the plan of being fossil fuel focused.

LETAG's composition was dominated by fossil fuel energy users and producers. They worked directly with the government to develop the energy plan. It was something that the government was not keen to publicize.

According to notes taken by one of the executives during a LETAG meeting, the Industry Minister Ian Macfarlane stressed the need for absolute confidentiality. The Minister saying that if the renewables industry found out there would be a huge outcry.

Until today Libby Anthony, Chief Executive Officer of the Wind Energy Association was not aware that this group existed and that such meetings had taken place.

LIBBY ANTHONY: Absolutely. The wind industry is disappointed that these meetings are going on and we don't have a seat at the table. It's disappointing that they have not involved the renewables industry in this process, maybe they don't understand the benefits and the opportunities of the wind industry in particular and the renewable energy in general.

ANDREW FOWLER: Though the government did meet with the renewables industry, they were never invited to join this key government advisory group process.

We have also obtained the minutes of a LETAG meeting during which Gary Wall, General Manager of the Energy Futures branch of Department of Industry stated that government was seeking to adjust policy so it supports and accommodates industry's direction. But Ian Macfarlane defended the composition of the advisory group.

IAN MACFARLANE: They were people who used, produced, or were involved in energy and I mean that takes in a broad spectrum. We were running a concurrent process with the renewable energy sector and both David Kemp and I were meeting with the renewable energy sector quite regularly. The government makes no apologies for consulting with industry right across the board.

ANDREW FOWLER: The wind energy producers say that they had no idea that the LETAG meeting was taking place.

IAN MACFARLANE: Well I'm not aware of whether they knew about it or not. The reality is that probably the LETAG group didn't know I was meeting with the wind or the solar people as well. I mean government has a process of meeting with all groups. Those meetings are commercial in confidence, things are discussed in there on the basis that people are open and frank with me, and I am open and frank with them.

MARK COLVIN: Industry Minister, Ian Macfarlane, ending that report from Andrew Fowler.

Source: Australian Broadcasting Corporation (broadcast on 7 September 2004)

Analysing the teams

When we look at issues we often think in the abstract: these are the strengths and weaknesses, these are the threats and opportunities. But in reality change always comes down to people. So who is on our side and who is not? How much influence do we have versus our opponents? How do you get key people to get off their bottoms and take an issue seriously? How do you excite a politician to take up a cause? Why will an individual write a letter to the newspaper?

In the previous section we looked at the potential stakeholders of the transition to renewable energy, the issues that might arise for them and their resource capacity to respond. It is these groups and the individuals behind them that become the pool of players from which both sides – the proponents and the opponents of renewable energy – can be drawn.

Definitely on the renewable energy (RE) team are the renewables industry, large environmental groups working on climate change or energy issues and parts of government.

Definitely on the other team are pure fossil fuel, nuclear or large hydro suppliers (those with no renewable energy interests), energy-intensive industries, and other parts of government.

Players who could go either way are energy companies, energy consumers, smaller environmental groups that do not work on climate change, single-interest environmental groups (birds, forests), unions, farmers, land and property holders, conservation groups, and local and state governments.

Sitting on the bench are the general public, finance sector, insurance sector, medical and health groups, churches, scientists, educators and business lobbies.

How do things look so far?

Well, the odds are not looking good for a renewable energy win as the teams run onto the field. Each team has two players. On one side the renewable energy sector is barely out of nappies in this new market (as we assume it has no helpful legislation in place), accompanied by the environmental movement. On the other side are two big and beefy industries, the suppliers of conventional energy that dominate the market and the energy-intensive industries that produce important goods like cars or steel.

Meanwhile, some parts of government are on the pitch because of energy security issues and the need to act on climate change, while other parts of government are there to show concern for the economy and industry, the health of which they may believe to be underpinned by access to cheap energy.

In the early days of this game there may be little conflict. The big energy companies may do a little dabbling in renewable energy, put some solar panels on a school, investigate a wind farm or two. However, once renewables start having aspirations about market share and legislation, the opponents' true colours may be seen. Some may be willing to adapt to the renewables power play, while others may resist.

The energy-intensive industries will be hard-pressed to make the best of anything that leads to an increase in power prices and therefore their product. If the changes occur equally to their competitors then they may be more comfortable, but if they see energy prices eroding their competitive position then it may be considered an issue of survival.

Perhaps worst of all, both these anti-renewable team members' interests overlap considerably. All members of these two opponent groups want lots of sales and cheap prices and they will often work closely together. For instance, in the Australian MRET debate, the Australian Aluminium Council, the Australian Coal Association and the Minerals Council of Australia provided joint submissions to government on renewable energy policy and commissioned expensive consultants to critique renewable energy market studies.

We need a plan!

If we watch this game play out without intervention, the most likely achievement will be a set of toothless policies that spawn some good-looking showcase renewables projects but fundamentally leave the system unchanged. That is, the renewable energy side will lose.

So the renewable side must have a game plan, and this will almost definitely require getting more powerful players onto the field in their support. There are clearly some excellent new players to look for, but we must bear in mind that the opposition may not sit idly by.

In the land of PR and political campaigning, the catchphrase for what we now need is 'the communication strategy'.

A Campaign Strategy for Renewable Energy

We must maximize our chances with decision-makers who have the power to legislate. To do so we must strive to maximize the decision-makers' room to make positive decisions and minimize their room to make negative decisions. Therein is the basis of a campaign.

When Greenpeace International started work on its highly successful campaign for offshore wind energy development in the North Sea, a very seasoned press officer handed me a piece of paper with ten questions on it. 'Answer these,' she said, 'and we've got our communication plan.' I have rearranged these questions slightly for our purposes.

1 PROJECT OBJECTIVES: What needs to be achieved?
2 TARGET AUDIENCES: Who do we need to reach?
3 SITUATION ANALYSIS: What is the noise above which we must be heard?
4 STRATEGY: How will we reach them?
5 KEY MESSAGES: What do we need to communicate?
6 TACTICS: What are we actually going to do?
7 SCHEDULE: How much time do we have and when will we do what?
8 BUDGET and RESOURCES: What human, financial and other resources are required?
9 MEASUREMENT CRITERIA: How do we measure our progress?
10 DYNAMICS: How do we keep abreast of change?

For people with communications training these points are part and parcel of their everyday work; and I do not want to do an inexpert step-through of communication theory. However, for those unfamiliar with this field it is important to know how and why such campaigns are constructed. For experts, I hope the discussion will provide some insight into the peculiarities of renewable energy communications strategy. In striving to provide practical illustrations of how these ideas and concepts have been given life in the real world I will largely rely on the AusWEA MRET campaign as a rolling case study, for the reasons described at the start of this chapter, but please refer to Chapters 6–11 for other examples.

In 2002 the Australian wind energy industry found itself in a strange position. The federal renewable policy driver, the MRET, had resulted in significant industry growth, albeit from an extremely small base. Polling commissioned a year later by AusWEA leads me to believe that Prime Minister John Howard's government felt compelled to introduce the legislation due to the high popularity of renewables among the public. (For example, 80 per cent of poll respondents felt the Prime Minister would do better to install more renewable energy than to sign the Kyoto Protocol; see Table 5.1.)

Yet the very low 2 per cent target set by the legislation was poised to effectively become a glass ceiling that would within years stall wind industry growth. Moreover, because the target had been translated into a flat target of 9500GWh by 2010 compared to 1997 levels, it would not actually increase the overall share of overall renewable energy generation, due to rising overall energy demand. Many groups, including AusWEA, viewed the 2003 review of the MRET legislation as a chance to increase renewable energy development.

At the same time, communities were raising concerns in the media about wind farms, and issues seen in other countries – birds, noise, landscape and property prices – were starting to emerge along with stories about wind farm objection. These were on a trajectory to overtake good-news stories. As I stated in a presentation to AusWEA at the time, wind power is an iconic clean, green image which has essentially become a brand. Wind farms are regularly used by environmental groups and socially responsible businesses as a symbol of a better future (see Figures 5.2 and 5.3). Yet no one has control or responsibility to protect and develop this common brand.

Overall, it was a dangerous situation for any industry and I was asked to put together a proposal to address some of these issues, and the resultant campaign was carried on throughout 2003 and into 2004 by myself and then-AusWEA Vice President, Mr Rick Maddox. The many battles won or lost along the way will provide useful illustrations for renewable energy industry campaigning.

Project objectives (What needs to be achieved?)

Defining objectives

Getting the objectives right is critical. The objectives we choose must lead to real and sustained increases in renewable energy generation. It is always tempting to suggest, 'Our objective is to raise a debate about ...' or 'Our objective is to raise the profile of ...'. However, these are not ends in themselves, rather they are tactics

Table 5.1 *A public opinion poll commissioned by AusWEA confirms public approval is high for actions that support renewable energy*

ACTION	LEVEL OF IMPACT ON PERFORMANCE PERCEPTION						
	Much less likely	Less likely	Makes no difference	More likely	Much more likely	Don't know	Mean
	(1)	(2)	(3)	(4)	(5)		
Increase the amount of electricity generated by non-polluting means such as solar or wind energy	<1%	1%	18%	53%	27%	1%	4.06
Increase research on reducing greenhouse emissions from coal-burning power plants	<1%	3%	30%	53%	12%	2%	3.76
Sign the Kyoto Protocol which is an international agreement to reduce greenhouse pollution	<1%	3%	34%	45%	14%	4%	3.73
Reduce land clearing	1%	3%	54%	29%	9%	4%	3.42

ELECTRICITY OPTION	LEVEL OF SUPPORT					
	Strongly oppose	Oppose	Support	Strongly support	Don't know	Mean
	(1)	(2)	(3)	(4)		
Building wind farms	1%	2%	27%	68%	1%	3.64
Building gas-fired power plants	11%	29%	43%	7%	10%	2.52
Building new coal-burning power plants	34%	39%	17%	4%	5%	1.90

Source: AusWEA.

to employ to achieve the real and measurable goals. In the case of renewables, I would argue that objectives tend to set themselves. Given the policy areas covered so far in this book, we now have a very clear idea of what works and what does not. Any campaign objectives must be in keeping with this understanding.

Although there may seem to be a continuous gradation of success, there is in fact a critical threshold to be crossed: either the policies create a self-sustaining environment in which renewables grow, or they do not. If the policies in place cross this threshold, we can be confident that a major objective has been reached. Until that point – and it may not be achievable in one go – our job remains incomplete and the ultimate objective is yet to be achieved.

In a one- or two-year campaign or within a given budget it may simply be impossible to move from a given starting point to see the required policies enacted. However, it may be possible to consolidate steps along the way. Perhaps you will witness legislation to allow consumers to purchase green power, or legislation to allow third-party access to the grid, and so forth. But as we will see, the assets

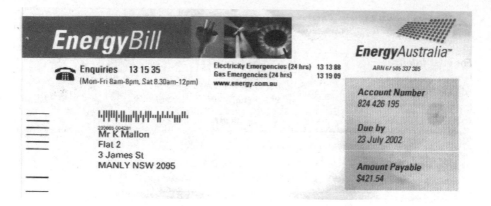

Note: The only form of generation shown is a wind turbine, for an actual energy mix that is 98% coal!

Figure 5.2 *Australian energy bill*

Reproduced with the kind permission of The Body Shop International plc

Figure 5.3 *Billboards seen in Body Shop stores around the world*

> **Box 5.2** *Defining objectives for the MRET campaign*
>
> The MRET campaign began with two main objectives. The first was to reverse growing negative perceptions of wind energy and foster strong support within wind project stakeholders and the wider community. The second was to secure the MRET and then achieve a significant increase in its target and its effectiveness to deliver an internationally competitive renewable energy industry.
>
> How did we arrive at these goals? At the time of preparing the campaign, two issues loomed large. The first was the review of the MRET legislation which created the possibility for an increase in the legislated target. The second was a measurable increase in the volume of objection to wind projects in the media.
>
> In fact the two objectives were inextricably linked. We needed to win over the public as opposed to leaving anti-wind sentiment unchecked, and we also needed the public to be seen by decision-makers as supporters of renewable energy. These two goals would in fact reinforce each other, because it is easier to engage people and get them to be supportive when their decisions to do so are connected to an action. The obvious cause for the public was to tangibly support the bid for an increase in the MRET target.
>
> Our second objective, which was initially to simply seek an increase in the MRET target, had to be expanded after a government review into energy industry restructuring; worryingly, it proposed to scrap the MRET scheme altogether in favour of carbon trading. This proposal reflected the availability of cheap short-term options for emissions reductions in the form of carbon sinks and energy efficiency. If the proposal to scrap MRET had been accepted, the carpet would have been pulled out from under the Australian renewables industry. Thus the second objective was updated to include the need to save MRET from extinction.

and allies typical to renewable proponents make it critical to engage the public. Therefore the objectives must touch members of the public in some way in order to build up the political capital of the industry and its proponents.

Target audiences (Who do we need to reach?)

With objectives now defined, it becomes necessary to deduce who must be influenced to achieve said objectives. As we have said, in a political campaign for renewable energy the final measure of success is legislative change. Therefore the decision-makers on policy – politicians – are the ultimate target audience. Nevertheless, things in politics rarely work linearly and there will be many individuals and groups along the way who feed into the path of influence of these decision-makers.

It is tempting to assume here that since renewables are great for saving the planet, if politicians only knew this they would definitely make the right decisions. Invariably, however, ministers are balancing many different considerations and external pressures, and above all are mindful of maintaining their position in power which entails a very different set of considerations. Thus, to affect decision-making at these levels, one must understand these politicians' situation and pressures and adapt strategies, messages and tactics accordingly.

The decision-makers

The key ministerial decision-makers are, of course, those with portfolios covering energy, the economy, environment and industry. All of these portfolios are crucial. Renewables obviously have environmental outcomes, and as labour-intensive industries create lots of jobs, renewable legislation can also affect the cost of energy production which in principle affects the entire economy.

Having identified the decision-makers, we can start to define them in relation to our cause. We can consider who they are and their areas of interest or identification. We can understand what the balance of power between portfolios looks like and the key concerns we will need to address. All of this will help us derive the information that we will need to develop and convey our key messages. However, we will also need to identify the obstacles that decision-makers will face if and when they do set about installing the required legislation. This will shape the campaigns that will need to be run by the various renewables proponents.

In a perfect world, renewable energy proponents would walk into the Prime Minister's office where the ministers of finance, industry, energy and environment would be assembled waiting for a thorough briefing. The renewable industry representatives would convince them of all aspects and present the perfect package of policy measures for success. These would be legislated forthwith and a bright and glorious future for renewables would soon blossom. However, although it is essential that the industry get in front of decision-makers (one would be surprised how often they do not) in fact advice from industry forms only part of the input those ministers consider. The next step is to know the other sources of their advice.

Who will the decision-makers listen to?

So who are these individuals or groups that lie in this mysterious path of influence? It is probably easiest to expand from the centre of influence outward. First we have influencers inside politics or the bureaucracy; second we have influencers from outside politics; and third we have the public.

INFLUENCERS FROM WITHIN POLITICS

Ministerial advisers and government agencies
Please note that while I will refer throughout to a parliamentary system of government, its structures and positions will have logical parallels in other systems of government.

Ministers solicit advice from advisers and other ministerial staff members as well as any government agencies of energy, environment, industry and economy. Because these people interpret and filter information for their ministers it is absolutely essential they have access to correct information and know the viewpoint of the renewable industry and civil society across the issues involved.

These positions are important and necessary; nevertheless people do come into them with their own views of the world. Sometimes this means that any message passed upwards comes with prejudices attached – positive or negative.

Of course these advisers and agencies may also experience lobbying pressure from opponents of renewable energy legislation.

Elected representatives

Government ministers agree and propose legislation, however, it will be a wider parliament that amends, discusses and ultimately votes on the legislation. This will include representatives from within the governing party and from other parties or independents. Clearly their views and opinions affect the formation of the legislation well before it is actually tabled in parliament. The balance of power and trade-off between issues can make the non-governing or minority parties very influential.

Finally and perhaps most importantly, we must repeat that decision-makers do not make decisions on scientific or political merit alone. While they are in principle reflecting the wishes of the public they are also ensuring that they stay in power.

This gives rise to a completely new set of influencers on the decision-maker. There is the issue of the decision-maker's own status within government and the standing of their government with the voting public. There may also be backbenchers (non-cabinet members of parliament) in marginal seats who are worried about losing their seats because of green voters. These politicians are ready to pressure decision-making ministers for initiatives that will help to prove the government's green credentials. There may be members from small parties who have bargaining power with the government over completely different legislation.

Furthermore, among these parliamentarians strong renewable energy supporters may be found. If we are really lucky, these individuals may even be members of the cabinet. There will also be cross-party committees considering and advising parliament on various issues. Often the individuals populating such committees are enthusiastic about the issues and therefore potentially supportive.

Political champions

Research reveals that the most educated and successful people in society have the greatest degree of environmental awareness – not quite the dreadlocks and bongos environmentalist stereotype often imagined. Therefore it is highly likely that among politicians there exist allies who can help, advise and convey messages upward. Better still, there may be champions among this pool: people prepared to act as political leaders in the campaign for legislative change. The higher up the champion in the system, the more power and weight they have to drive the ball toward the goal. One such champion was Svend Auken, the Danish energy and environment minister who maintained legislative support sufficient to transform Danish wind manufacturers into a major economic powerhouse (see Figure 5.4).

Seeking out champions is an important part of a campaign. Once they are found, a campaign should not fail to support and nurture these relationships as it all ultimately comes down to people. However, a champion can only help lead a campaign for change, not run the entire campaign. The role of the wider campaign is to build the external pressure for change and then create space for the champion to run with the ball. The heavier the champion is politically, the more they will be able to do alone; the lighter they are, the more work the campaign must do to clear the path for them and protect them from being brought down.

Source: Greenpeace International

Figure 5.4 *Political champions: Danish Energy Minister Svend Auken and then UK Environment Minister Michael Meacher lend support to offshore wind energy aboard the vessel MV Greenpeace as it departs for the Turno Knob offshore wind farm*

Again we are maximizing room for positive decisions and minimizing the chance for negative decisions.

INFLUENCERS OUTSIDE POLITICS

At this chapter's outset, we pointed out that the team of renewable energy proponents may start as only the (still small) renewable energy industry, the larger environment groups and some government agencies. This is hardly a heavyweight package with which to convince governments to adopt major energy sector reforms. These voices may also not be the source from which certain messages need to be heard; is an environmental group seen to be a credible commentator on what is best for the economy? So we need to discover the individuals or groups that decisions-makers and their direct influencers heed.

Where might we find these people? There will always be individuals or groups that have a trusted place in the hearts of decision-makers or perhaps have influence for other reasons. These people and organizations will help us influence decision-making if we can convince them of the merits and arguments in favour of renewable energy. With issues like climate change, there will be people and organizations that may be concerned, but have not yet spoken out on an issue – doctors and health professionals, churches and international aid groups, scientists and academics, industry groups and unions, farmers and innovators, large companies or popular retail brands, or the finance sector and insurance companies. We can see, of course, that there is a good deal of overlap between the influencers we are identifying here and the stakeholders identified in the previous chapter.

The influence of such groups or individuals is not just upward. They can also have considerable influence in wider society and with the media. For example, if the head of a big oil company speaks out about the risks of climate change and the need to support renewable energy, he will be reported in the business pages rather than the environment pages, and will reach and perhaps sway a very different audience. The British Wind Energy Association's 'Embrace Wind' campaign has used high profile 'stars' to champion wind energy.

On the other side of the coin are individuals and organizations that can be extremely influential if they oppose renewable energy legislation. Large energy companies have very high-level and direct access to ministers and can convey messages that will not be heard elsewhere. It is therefore necessary to include in the target audiences those groups that may oppose renewable energy legislation.

THE PUBLIC

Given the lack of significant influence on the pro-renewable side of the equation at the campaign outset, raising public awareness on the issues and enrolling their vocal support for safe energy represents a critical opportunity to swing the process in our favour. Government decisions will not be made on merit alone. Not only will there be high-level influence through lobbying, there will also be local-level exposure to the voting public.

Ultimately the power of elected representatives rests with the voting public. We may, of course, be looking to start the industry in new markets where countries may lack a conventional democracy. Yet politics are politics whatever the structure. The need to build up public support for renewable energy within the wider community and its stakeholders will always exist. Even if fantastic legislation were in place, there would still be a need to appeal to the public to accept the changes brought by renewable energy. This appeal also extends to organizations that the public looks to for guidance.

However, it is often said that there is no such thing as the general public. The critical publics to any renewable energy campaign will be stakeholders and the politically important parts of society. Stakeholders were discussed in detail in the previous chapter, so now let us examine the important parts of society.

Obviously the form taken by influential parts of society varies from country to country and also with the political party in power. Sustainable energy issues have the fortune of being generally non-partisan and having broad appeal. Nevertheless,political parties have antennae tuned to certain demographics. It may be the middle ground of people with young families, the youth vote or the rural vote. Many countries are now witnessing the rise of a green vote which leaches votes from both left- and right-wing political parties.

Table 5.2 *Support for renewables in Australia*

'In the year 2000 the Howard government set a target to increase the contribution of clean energy from renewable sources such as wind and solar over the next 10 years. Do you think this was a good or bad initiative?'

		Very bad		Bad		In between		Good		Very good		Don't know		Mean
		1		2		3		4		5				
Total		7	1%	22	2%	17	2%	332	32%	636	62%	13	1%	4.55
Gender	Male	4	1%	12	2%	8	2%	161	32%	317	63%	4	1%	4.54
	Female	3	1%	10	2%	9	2%	171	33%	319	61%	9	2%	4.55
Age group	18–29	1	1%	6	5%	3	2%	46	35%	73	56%	1	1%	4.43
	30–39	2	1%	4	2%	3	1%	69	34%	125	61%	2	1%	4.53
	40–49	1	0%	4	2%	3	1%	65	32%	127	63%	2	1%	4.57
	50–59	0	0%	2	1%	3	1%	68	31%	139	64%	4	2%	4.62
	60+	3	1%	6	2%	5	2%	84	31%	169	62%	4	1%	4.54
Household income	<$20,000	4	2%	6	3%	5	2%	84	35%	132	56%	6	3%	4.45
	$20–40,000	1	0%	8	4%	2	1%	68	31%	135	63%	2	1%	4.53
	$40–60,000	0	0%	3	1%	1	0%	65	32%	132	66%	0	0%	4.62
	$60–80,000	0	0%	2	2%	2	2%	39	33%	74	62%	2	2%	4.58
	$80–100,000	0	0%	2	3%	0	0%	17	26%	45	69%	1	2%	4.64
	$100,000+	0	0%	0	0%	1	1%	33	34%	63	65%	0	0%	4.64
Voting intention	Labor	2	1%	5	3%	5	3%	61	36%	97	57%	1	1%	4.45
	Liberal	1	0%	5	2%	3	1%	93	29%	220	68%	3	1%	4.63
	National	0	0%	1	6%	0	0%	5	29%	11	65%	0	0%	4.53
	The Greens	0	0%	1	1%	0	0%	16	22%	55	75%	1	1%	4.74
	The Democrats	0	0%	0	0%	0	0%	5	28%	13	72%	0	0%	4.72
	An independent	1	3%	1	3%	0	0%	10	32%	19	61%	0	0%	4.45
	Other	0	0%	1	7%	0	0%	3	20%	10	67%	1	7%	4.57
	Don't know	3	1%	6	2%	7	2%	118	35%	192	58%	7	2%	4.50
Children	No children	0	0%	4	3%	3	2%	47	36%	76	58%	0	0%	4.50
	Intend to have children	1	1%	2	2%	0	0%	38	32%	77	64%	2	2%	4.59
	Already have children	6	1%	16	2%	14	2%	245	32%	479	62%	10	1%	4.55
	Children at school	2	1%	8	3%	4	1%	108	36%	176	58%	3	1%	4.50
Location	City	2	0%	9	2%	6	1%	154	29%	351	66%	6	1%	4.61
	Country/regional	5	1%	13	3%	11	2%	178	36%	285	57%	7	1%	4.47
State	NSW (inc. ACT)	1	0%	13	5%	2	1%	91	36%	142	56%	6	2%	4.46
	Vic	4	2%	3	1%	4	2%	62	30%	130	63%	3	1%	4.53
	QLD	1	1%	4	3%	1	1%	48	32%	98	64%	0	0%	4.57
	WA	0	0%	0	0%	3	3%	33	32%	67	64%	1	1%	4.62
	SA (inc. NT)	0	0%	0	0%	5	2%	60	28%	145	69%	1	0%	4.67
	Tas	1	1%	3	3%	2	2%	38	38%	54	54%	2	2%	4.44

Note: Cross-tabs on support for renewables across demographics can provide very useful insight into critical parts of public opinion. In fact support for renewable energy is surprisingly uniform.
Source: AusWEA (2004); personal communication

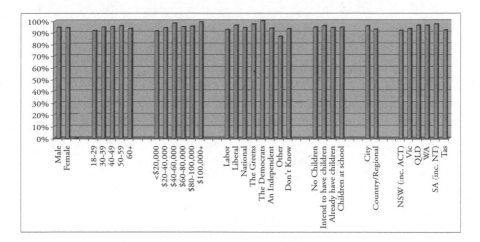

Source: AusWEA (2004)

Figure 5.5 *Uniformity of support for renewable policy over 33 demographic types*

Situation analysis (What is the background noise above which we need to be heard?)

If our target audience is to hear and understand our key messages then we must make sure our messages resonate for that group. Some messages and issues will naturally rise to the top to dominate media and public space (such as elections and wars) while others must jostle for space in competition with everything else that is going on. The chance of our messages being received is enhanced if we choose opportune moments that are not dominated by other issues, or if we find ways to reinforce our message with the other issues in circulation. Simply put, we are trying to build a map of the media and issues landscape over which the communication plan must lay.

A friend of mine who is a very fine campaign strategist recently applied for a job with Amnesty International. She was being interviewed by telephone in the early hours of her morning at the time of the US-led invasion of Iraq when asked, 'Given the war in Iraq, how do you think Amnesty could best manage its campaign about prisoners of conscience?' Her reply was, 'Tell the staff to take their annual leave.' She got the job.

We need to look at the situation as it affects our target audiences. We can split this situation analysis into three parts. First there is what is going on more generally (for example, the focus in the national media) which we will call the *external situation*. Second is the current focus of discussion in our sector, the energy or environment sector, which we will call the *internal situation*. And third are the situations of each of our target audiences, the *target situations*.

Box 5.3 *The various MRET publics*

During the MRET campaign, we noted that there were several social groupings among the public whose opinions we felt would resonate with the government, and for whom we felt we had a good-news message to convey. These were:

- middle Australia, namely working people with children at home;
- people from the business and finance sector; and
- rural people.

This illustrates how it might be useful to separate the general public into specific publics with particular needs, concerns and expectations. This enables campaign strategists to tailor and focus the key messages (and media) accordingly.

The external situation

Obviously the media and the buzz on the street affect how far and how quickly messages can get out. If the campaign involves the public and the media then these issues are critical.

In great campaigns, the campaign itself becomes the noise, the centre stage of news, opinion and discussion. This has happened time and again with environmental 'stop' campaigns. However, in my experience I have never witnessed this with solutions-oriented work. The closest this level of buzz has come to centre stage for renewables may have been the protest marches in Germany to save the feed-in law. In general solutions are good news and good news is rarely big news. However, by saying 'STOP blocking renewable energy!' the message revives the conflict we were missing.

Having said that, sometimes other news can provide a platform for entry. Perhaps the discussion about oil shortages and high prices, or water shortages and the fact that coal plants use up to 2 litres of water per kilowatt hour, climate change stories and so on.

The internal situation

The internal situation analysis at its most simple starts with a statement of the status quo. Who stands where on what issue? How significant is our issue compared to other energy or environment issues? How significant is it in the minds of decision-makers?

We can also consider the energy debate and coming issues that might trigger new debates, such as blackouts and fuel price increases. Where does the environment debate stand? Are climate impacts such as heat waves or floods likely to set off new rounds of conjecture? Is anyone releasing reports or having a meeting that will prick the interest of the media?

As renewables gain ground and exposure we see the rise of groups opposed to renewable energy developments in many countries. Although their concern may be about a single project, aggressive tactics and factual deception can lead to campaigns which smear an entire industry. In some cases support for such groups has been linked to competing industries, but one must be wary of overly conspiratorial

Box 5.4 *Use of current issues to leverage media coverage*

Wendy Frew

The Carr government's decision last month to clear the way for more coal-fired power plants will exacerbate NSW's water shortage and could further damage the state's struggling river system, according to a briefing paper issued by conservation group WWF Australia.

Coal-fired plants are big water users, competing for water for domestic use and farm irrigation, according to WWF, which has called for a moratorium on building more coal power stations.

'NSW is already highly vulnerable because of this incredible dependence on coal-fired electricity,' said WWF's climate change manager, Anna Reynolds.

'It just doesn't fit with a modern well-planned state to have such a high dependence on a fuel that is polluting and drawing incredible water resources when the state is in a water crisis.'

NSW relies almost solely on coal for its electricity generation.

A draft of the government's energy White Paper, obtained by the *Herald*, revealed plans to tackle rising demand for electricity, mostly due to the popularity of air-conditioners, by upgrading plants and leaving the door open to the private sector to build more coal plants.

Electricity consumption is forecast to grow at 2.2 per cent a year, with peak demand (when demand surges because of very high or very low temperatures) growing by 2.9 per cent a year.

The state's four heaviest freshwater users – the Mount Piper, Wallerawang, Bayswater and Liddell coal plants – use about 84,000 megalitres of water a year – equal to a fifth of Sydney's residential water needs. They draw the water from the already struggling Cox's and Hunter rivers.

WWF said generating one kilowatt hour of electricity from coal requires more than 1.5 litres of water: 'For example, to generate electricity for the average home from such a power station requires the use of about 20,000 litres of fresh water each year.'

Renewable energy options such as wind and solar use no water, while gas power stations, used for non-peak demand, use about half as much water per unit of electricity as coal.

The Utilities Minister, Frank Sartor, told the *Herald* efforts were being made to address power plants' use of water.

Source: Sydney Morning Herald (2005)

Box 5.5 *Selected text from a perceptions audit performed by Republic consulting for the AusWEA MRET campaign presented at AusWIND 2003*

Introduction

- Republic conducted perceptions audit with over 12 key players
- Specifically we investigated
  - Positioning of wind energy in the renewables debate;
  - Policy and political imperatives driving energy sector reform;
  - Opportunities to advance the case for wind energy.
- Republic has summarized the feedback received and provided recommendations to further AusWEA's objectives

The importance of the coal industry must be acknowledged... but neutralized

- Considerable scope to minimize coal industry impact
  - Little recognition that it is no longer a zero sum game
    - "If your expansion has to be at the expense of coal it's not on."
    - "The coal industry sees renewables as a real threat. They will do everything they can to protect their market share."
  - Economic benefits to regional areas holding great sway
    - "Convince the government that wind can generate jobs in the bush without harming coal and you're home and hosed."
    - "Getting support for wind energy development at the community level will be critical."

The industry must continue to build the business case for wind energy

- Decision-making process is being driven by political imperatives as much as policy ones
  - Coal leverage
  - Cost burdens
  - Wind gaining traction as bush solution
    - "Wind wasn't even on the radar screen 12 months ago but now there is growing recognition of real potential. They need to run with that."
- Lack of clarity over time required for renewables to become competitive
  - "They're asking for an open ended subsidy with no sunset clause. We need to see our exit strategy."
  - Estimates of cost convergence

A dual positioning campaign can be adopted to pursue broad-based support

- Third parties and advocates must be identified
  - Supporters of 10x10 position
  - Listened to by governments
- Two lines of argument adopted:
  - Environmental positioning
    - Little sway with feds but helpful with states
    - Senate hearings and cross-bench support
    - Garner third-party support
  - Economic Impact
    - Key regional electorates
    - Grassroots effects
    - Backbenchers and key supporters

Source: Republic (2003)

---

attributions for these movements; competitors may stir the pot, but they need fertile ground. It seems increasingly apparent that such opposition may be part of the broader context and require preparation in any renewable energy plan. These issues are mentioned in this book in Chapter 9 on Spanish renewables policy and are worthy of quite specific consideration.

Target situations

How do decision makers view our issue? What is occupying their attention now? What are the possible links or reinforcements that can bring our issue to their attention? What about each of our other target audiences?

Clearly we need to know what can make our messages resonate. For instance, farmers might be concerned about the prospects of energy crops or the threat of drought. For builders it may be concerns about added costs of PV versus ability to add value.

Our target situation analysis must also incorporate how positions evolve as a result of the campaign. For instance, we noted industries and organizations that may oppose renewable energy legislation. As the issues increase in profile we can expect this particular target situation to evolve and become more acute for these groups, who will respond accordingly.

Analysis tools

Now that we know the situations we want to analyse, how do we actually do that analysis? There are many tools available and much will be obvious to those in the renewables industry if they simply read the papers. Some other useful tools include media analysis in which media monitoring companies conduct specific reviews to gather information.

Perceptions audits are also useful (see Box 5.5). Here a PR company will call approximately 20 influential people from a given target group, and these individuals will agree to speak anonymously about an issue. Focus groups are another tool, similar to a perceptions audit but representing the public (or a subsection of the public) through selection of a group of people who discuss in detail their views and give reactions to messages or statements. Another tool is polling. This is a more broad-brush approach that can be used internally and/or externally as part of a campaign.

Strategy (How will we reach the audiences?)

Strategy links tactics (what we actually do) to our objectives (what we intend to achieve). So far we have defined our objectives, worked out who can implement the legislation that will deliver the objectives, and determined who lies in the path of influence. We have conducted a situation analysis to map out the field upon which all the players move. Using this map we must now develop our strategy – the game plan. It is very important to go through this strategic exercise; although stepping straight into doing things can be very tempting, it is a fast track to failure.

I mentioned earlier that our basic strategy must be to maximize the chances of a positive decision and minimize the prospects of a negative one. Our campaign strategy works entirely towards these two outcomes. Another way to think of this interaction is to use the idea of political space in which we map the space available to a decision-maker on a given issue.

I will next discuss theory and strategy, and use the following section on implementation and tactics to provide concrete demonstrations.

The concept of political space

A decision-maker does not have an infinite range of options from which to choose legislation. Rather he or she is making decisions which balance sets of more limited options. A politician may have to balance technical recommendations regarding a particular issue with competing interests from other portfolios such as economics and industry. Political pressure may be applied from within their party from particular electorates, or from other parties. There is also pressure from society and the media. We can picture all these forces as creating or diminishing the room in which the decision-maker can move on a given issue.

Our strategy must therefore block out negative decision space, secure positive decision space, and re-open positive decision space where it may have been closed off.

Let's run through a scenario to help move this out of the abstract. Our player wants to take his precious legislative package (the ball) all the way up the field across the legislation line. But a big obstacle is set in the way, labelled 'renewable energy is expensive so the economy will suffer'. Another obstacle, 'the consumer will have to pay more!' is also blocking the path. Meanwhile, the environmental groups have put in a backstop: 'climate change is on the public agenda and inaction is not an option' (securing positive space). However, this leaves open the option of carbon trading rather than growing renewables, so how do we address this one? To get past our obstacles we must open up a path. And we may need to call upon stronger arguments or more influential allies to push the obstacles and opponents aside.

So in this simple case, our strategy might include the need to demonstrate that cost increases are small, or that consumers are willing to pay any increase (re-open closed space). We also need to show that carbon trading without renewable development will be more expensive to the economy in the long run (block out negative decision space). We may choose a strategy of using politically powerful job creation in rural areas to push aside arguments about detrimental impacts to the economy (secure positive space).

From this will drop out the tactics that we must employ to implement the strategy. For example, our tactic may be commissioning a report on the cost of our legislative proposals or a poll to ask whether consumers would be prepared to pay for the increase in clean energy.

The lesser of two evils

Decision-makers are often in the unenviable situation of being damned if they do and damned if they don't, or to put it another way, the space for decisions is blocked in both directions. In these situations, political circumstances often take precedence, meaning the least politically damaging path is selected. Most people are in politics to make a difference and create positive change, and in order to do this they have to maintain political support from their colleagues, their opponents and voters. If they undermine this significantly in one particular area it in turn undermines their general ability to do what they are there to do.

Therefore, support among the public for environmental protection and renewable energy is one of the renewable energy sector's greatest assets. The use of public – and therefore voter – sentiment will always be crucial if politicians are to be persuaded to make decisions to which there is known and potentially powerful opposition. Deciding how best to use this asset is a key campaign challenge that will be discussed under tactics.

Situation analysis and SWOT

To actually lay out our map and game plan, we must make sure that first, our situation analysis is fully represented in our mapping and second, that the potential changes to the map are identified so that we avoid as many surprises as possible. We can combine both in a SWOT (strengths, weaknesses, opportunities and threats) analysis, as shown in Table 5.3.

Top down, bottom up and middle out

With our strategy becoming mapped out above, we have an idea of what we need to do and where. However, to apply this strategy we must look to who can do it and how. For instance, academic reports may demonstrate that renewables can create six times more manufacturing and installation jobs than coal. However, this

Table 5.3 *SWOT elements any renewable policy campaign will need to consider*

STRENGTHS

- There are pressures for change in most countries which may include environment, energy security and employment issues.
- Environmental groups are usually strong advocates of clean energy and can have significant social influence.
- The renewable industry comes with little baggage so there are few reasons other organizations would fail to endorse it.
- Renewable energy has a very low threshold above which local manufacture can start.
- Renewable energy has a very high employment creation ratio compared to other technologies.
- Many renewable energies including biomass, wind and small hydro are focused in rural areas, meaning that investment and employment creation will be similarly focused.
- Many renewable energies are high-tech but easily transferred to new markets.
- Renewable energy is now mainstream in several markets and is considered proven or mature.
- Renewable energy often receives considerable attention from educators.

OPPORTUNITIES

- Climate change has wide and significant impact meaning that a wide range of pro-stakeholders can be accessed.
- Some companies with renewable energy interests have a high profile and significant commercial influence if they become engaged.
- High popularity of clean energy and environmental action may make it politically attractive to politicians.
- High levels of membership for social and environmental groups can provide access to significant numbers of individuals to engage in the issues.
- Global industry is large and some sectors have meaningful resources for market development that can be brought in from overseas

WEAKNESSES

- The renewables industry is usually very small compared to existing generators, so it doesn't possess the same economic gravitas for its political demands.
- The financial communities regard renewable industries as high risk in new markets. This can colour the view of decision-makers.
- Renewable energy costs are higher than conventional energy if externalities are not factored in.
- Variability of supply of many renewables leaves network managers nervous.
- Industry's image is dependent on all projects; thus one bad project for a given technology can give the whole industry a bad name.
- Industry is vulnerable to mistakes made in other countries.
- Renewables can be seen as a side issue or non-critical.

THREATS

- The threatened industries are big, wealthy and influential.
- Threats to big industries are perceived as threats to the economy and employment.
- Low energy prices are considered key to economic growth.
- Increases in energy prices lower competitiveness.
- Promises of future decrease in price can lead to arguments that it is better to delay implementation of renewable policy until it is cheaper.
- Good-news stories are always weaker than bad-news stories.
- Social groups can find it more difficult to rally around 'go' issues than 'stop' issues.
- Environmental and social supporters may not be seen as credible by decision-makers.
- The local environmental issues may have alienated potential allies.
- Actual projects can give rise to localized concern by green or conservation groups, for example, due to transportation of biomass or effects of wind turbines on the landscape. This means renewables can lose the moral high ground on environmental issues and create green-versus-green real or perceived confrontations.
- Anti-renewables groups can access willing support from overseas groups.
- Weak legislation can be used to defuse public concern, with the public unable to see that it is weak.
- Other environment initiatives – e.g. forests or wildlife – can be used to break the unified voice and get environment groups 'off the scent'.

fact will not secure political space unless it is known by various parts of the target audiences. For example, unions need to know that renewable energy manufacture and installation is labour intensive, so that the issue of a threat to employment cannot become a reason to decline support for renewables.

Given our three levels of target group (influencers inside politics, influencers outside politics and the public), there are three mutually reinforcing approaches we can take to directly influence and define the political space around the decision-makers. First, we can go straight to the decision-makers or their advisers. Second, we can enrol the aid of people and groups with weight and influence in the key areas our strategy has identified. And third, we can engage and build support from appropriate subsections of the public.

As for strategic pathways, for any renewable energy campaign there are many possible ways to proceed. One might be to build positive messages and address negative issues through materials and direct meetings with decision-makers, advisers and agencies. It may be beneficial to build positives and address negatives with influential third parties such as the finance and business sector or union movement.

Another strategic pathway entails creating a common platform with existing allies, then educating and enrolling other natural allies, and finally working with the mixed-impact stakeholders to address their concerns and build mutual support. This ally base can be used to spread support both upward into politics and downward to grassroots through the membership bases of the allies.

We can also take the strategic pathway of engaging certain publics for support and finding ways for them to act, and be seen to act, on behalf of the campaign cause. We can also address issues of public concern by addressing the reasons for the concern rather than the manifestation of the concern.

Key messages (What do we need to communicate?)

When you have positive information to convey, it is always tempting to try to say as much as possible in the belief that a person's opinion will be swung by the sheer weight of the good news. Sadly communication does not work like that.

I was once told that people retain about 3 per cent of the detail in a presentation and about 10 per cent of the general orientation. But mainly what they remember are impressions of the presenter. This is reinforced by what communications experts say about presentations to intelligent but inexpert audiences (like the media). To be retained, the information has to be fully comprehensible to an 11-year-old!

The bottom line is that complex information does not get through. Only one or two messages can be used with a given audience and these must be simple and quick to comprehend. These messages must also be reinforced over and over again!

The need for a common message

To find its mark, a message must be repeated again and again and preferably reinforced by repetition from as wide a group of influencers as possible. We must also

---

Box 5.6 *Media response as one arm of a communications strategy*

An organized and proactive media strategy can help to ensure that the positive side of renewable energy is heard, and also make it easier to respond when issues are first raised in the media.

By way of an example, consider the following strategy to leverage as much positive media space as possible: start by monitoring print, radio and television media, using a media monitoring service which provides coverage on topics you select; analyse the key issues and note the key commentators cited in the media reports; also note the journalists who are following and writing on these issues.

At the same time, build up a media database with contacts for the key journalists and media outlets, and ensure these contacts receive any media releases that are sent out. Prepare briefings on key issues that have been raised and proactively disseminate these briefings to journalists. Additionally, establish a system to react to the media on issues with letters or opinion pieces for print media, or verbal response for broadcast media. It will be important to react quickly – usually within a half-day for major media outlets such as large national daily newspapers.

---

bear in mind that many proponents may exist, each with very different reasons to promote renewable energy. Therefore, we cannot expect a single set of coordinated campaign messages.

There are two levels of commonality that we need to think about. On the one hand it is important that all proponents of renewable energy push in the same direction, even if they come from very different starting points. They may have different 'why' messages, but the 'what we want' message must be the same.

Within an industry or industries, however, it is more important that the 'why' messages are aligned and reinforcing, with no differences of view that can be exploited by renewable energy opponents.

Box 5.7 *Securing the European Union Renewable Energy Directive*

The need for a harmonized position cannot be underestimated. A disparate set of messages, even if they are not mutually exclusive can be used by decision-makers and bureaucrats as an excuse for inaction. I personally witnessed this as the European Commission procrastinated on drawing up the 'Renewable Energy Directive' which was expected to spring from the 'White Paper on Renewable Energy'. The Commission functionaries told me that they could not arrive at a draft directive because the policy demands of the different renewable energy lobby groups were completely different. They seemed content to use this as a justification for stalling the process.

However, all parties did agree that they wanted the Commission to promote more renewable energy in the EU. They also agreed that significant barriers opposed this aim, barriers which the Commission had to address. Over the course of about two months the major renewable energy lobbies and their environmental sector proponents negotiated a joint, mutually acceptable text. Their solution was quite simply to identify all the areas that the various parties agreed on, and as for the specifics of policy mechanisms they agreed to disagree! The text proposed the use of subsidiarity – an oft-employed term in the EU which means each country applies policies to achieve an objective as it sees fit.

With the excuse of irreconcilable demands removed, the process for a directive was back on the rails and ultimately passed into legislation (based on the principle of subsidiarity).

Below is the first page of the industry and NGO unified set of principles. It demonstrated that first, all key players in support of renewable energy were now in agreement on the basis of suitable legislation and second, they all wanted a legislated increase in the target for renewable energy in Europe.

Principles for a Renewable Energy Directive in the European Union

April 1999

In the European Union at present the electricity market is significantly distorted to the detriment of renewable energy generators: access to grids is restricted; excessive transmission costs are applied to renewables; there is still no internalization of environmental and social costs; embedded generators do not receive remuneration for the savings they create. These factors and the use of nearly 15 billion ECU[1] in direct subsidies to the conventional generation sector all contribute a market distortion that continues to hold back the harnessing of renewable energy in the European Union.

The signatories herewith believe that a European Union Directive, based on the ten principles presented, is required in order to redress the serious market imbalance and to establish a process for the orderly phase-in of renewable energy, together with the industry, jobs and climate protection that this will provide.

1   *A directive is called for, with legally binding minimum targets for each Member State, to promote the accelerated take-up of renewably generated electricity in the EU.*

2   *The directive must allow the most appropriate mechanisms for the delivery of the minimum target to be chosen by the individual Member State (subsidiarity) until such time as effective mechanisms have been proven by actual delivery of the renewable energy percentage minimum target.*

Choosing a key message

Whatever your own opinion, it is no prescription for key messages in a campaign. They may indeed evolve during the campaign's course. However, answering the following questions should help to filter out messages that will and will not work.

- Are the messages simple and easily comprehended?
- Will the messages address or provide a compelling balance to the obstacles in the mind of the decision-makers?
- Will they resonate and engage the target audiences?
- Will they provide a universal platform that individuals, groups, influencers and decision-makers can adopt as their own?

Box 5.8 *The GMO debate*

It is interesting to note that the genetically modified organism (GMO) debate in Europe has focused on health. Thus the target group one would infer is the consumer. The message has not been that genetically modified (GM) foods are dangerous, only that they *may be* dangerous. This is a simple and universal concept, and an easily adopted position for any third party as it is an almost impossible proposition to refute for any new product. As for the decision-makers, matters that affect health and safety override almost all other considerations, which severely limits their room to manoeuvre.

In Africa, by contrast, the key GMO issues are about seed control. The target audience has been farmers and the message has been that if you plant GMOs you will lose control of the seed source and thereafter your ability to decide what happens on your own farm. Two very different messages, for very different target audiences, from two quite different campaigns – but both with the same objective of curbing GMO production.

Messages to express why we want renewable energy

For renewable energy, there are a significant number of positive aspects from which we can select our messaging. The strengths listed in the SWOT analysis in Table 5.3 are elaborated below to underline the reasons why we must proceed with renewable energy development.

1   *Renewable energy is proven.* Replacing fossil fuel use with proven renewable energy technology and resources can safely achieve the deep (60–80 per cent) cuts in GHG emissions that scientists say are required to address climate change.
2   *Renewable technologies are mature.* Many renewable energy technologies such as wind, solar, small hydro, geothermal and biomass for power and fuels are now technically and commercially proven. Machinery has equivalent lifetimes, availability and reliability to non-renewable generation.
3   *Renewables foster energy security.* Appropriate renewable resources are often abundant in a country and can supply major portions of the energy demand if harnessed.

4   *Renewables are benign.* In the main, renewable resources and their harvesting are considerably more environmentally and socially benign than fossil fuel and nuclear generation. (However, no two projects or technologies are the same and this cannot be taken for granted.)

5   *It is a growth industry.* Renewable technology is cutting edge but it is also often easily transferred for manufacture in-country provided an adequate market is established.

6   *Renewables foster investment.* Local components and services for renewables create inward investment and can mean that significantly more of the internal investment in energy is spent domestically compared to other sources.

7   *Renewables create employment.* Renewable energy technology is labour intensive, especially during the construction and installation phases, and also in the operation and maintenance phase for some sources. These technologies create several times the amount of employment of equivalent fossil fuel plants.

8   *The cost of renewables is falling.* Renewable costs have declined radically and continue to fall. In some markets renewables are competitive with (new) conventional coal, nuclear and large hydro generation even before including the costs of health and environmental externalities.

Box 5.9 *Some examples of key messages*

The following texts are examples of messaging used during the 2003–4 campaign to increase the MRET legislation in Australia.

First, an excerpt from a media release that launched an AusWEA report highlighting the economic benefits wind power would bring to the nation:

> *The Driving Investment, Generating Jobs report launched in Canberra today by the Australian Wind Energy Association shows regional and rural Australia is already reaping significant employment and financial-benefits from wind power. Produced by energy specialist Dr Robert Passey, MSc, PhD, the report reveals setting the MRET at 10 per cent by 2010 would deliver total investment of nearly AU$7 billion including:*
>
> * *direct capital investment in Australia of AU$5.4 billion;*
> * *additional operation and maintenance expenditure of AU$210 million;*
> * *3500 additional manufacturing and construction jobs plus 280 additional operation and maintenance jobs …*

The following excerpt is from a media release that launched a report on the price convergence of wind power and conventional coal power. It is titled: 'Wind will challenge the cost of fossil fuels before 2020: Industry demands bigger market for renewables'.

> *Melbourne: The Australian Wind Energy Association (AusWEA) released a report today which indicates that Australian wind power will rival fossil fuel costs before 2020. However, the Association has stated that this will happen only if the industry is given room to grow through an increase in the Mandatory Renewable Energy Target (MRET).*

Source: AusWEA (available at www.thewind.info)

Figure 5.6 10×10 Briefing, *a briefing prepared for the Australian Wind Energy Association by Transition Institute and Rick Maddox*

Messages to express what we want

Obviously the chosen messages must also contain the 'ask' – the change required to realize the potential of renewable energy. These asks must obviously flow from the policy package being requested. We may have a long list of possible policy requests, the nature of which we have identified in detail earlier in the book.

Yet from this comprehensive policy package we must distil key messages that embody the essence of what will convince decision-makers and stakeholders. In the MRET campaign a detailed package included revised targets, timelines, baselines, phase-ins and penalty prices, which all became 'a 10 per cent target for new renewable energy in 2010'. This in turn became the even more simple '10×10' catchphrase (pronounced 'ten by ten') for the entire campaign.

Continuing with the MRET example, the key messages that we arrived at were as follows:

- Wind power is a low-impact way to help protect future generations from climate change.
- Wind power in Australia is a success – creating jobs, building factories and leveraging billions in investment. (The business sector is behind wind power, there are visible industry and employment outcomes and the governments can claim success.)
- The activity and investment is focused in rural areas. (The delivery is in some important areas where investment and employment are difficult to achieve.)
- Ongoing success depends on MRET being increased.

Note that these messages correlate directly to our target audiences – working people with children at home, the business and finance sector, and people living in rural areas.

It may surprise some that the AusWEA campaign did not excessively focus on climate change – which is the wind industry point of entry. There are two good reasons for this. First, it was already being done by the big national and international environmental groups and second, industry is not the best source for this message. In fact, some surveys in Europe indicate people trust the word of environmental NGOs more than governments, corporations and scientists. Thus companies and industry organizations are clearly not the most credible commentators on environmental or social issues: 'Well they would say that, wouldn't they? It's just so that they make more money.' So, if it is money and economic talk that people expect from industry, and if that is indeed their expertise, then better to talk that talk. Consequently the messages were focused on investment, jobs and factories and the politically important rural locations where this activity was occurring.

Implementation and tactics (What are we actually going to do?)

Now we are ready to implement the strategy. We know what we need to do and why, who we need to speak to and what we need to say. The next step is to actually build the roll-out plan for the strategy. If we need to get the message through to a particular person, we need a way to be heard, understood and have our message taken on board, and for the message to lead to action on their part.

There is always a temptation with campaigns to jump straight to tactics. Perhaps this is because tactics are easier to come up with than anything else. However, if the tactic fails to deliver the strategy, it is a waste time.

Most communication tactics are obvious but, in a world overloaded with information, it is often the more innovative ideas that find their target. Sometimes the best way to come up with implementation tactics is to put some clever people into a room with free drinks!

Tactics are very much about doing, so I've set out a selection of examples which can help make the point and give some food for thought for anyone trying the same, whether from industry, NGOs or government. Sometimes it is also useful to look at what works for other campaigns in a given country. If all else fails, one can go straight to the horse's mouth and find friends within the target group to ask what sort of vehicles work for them.

Tactics to maximize political space

Here is an example of blocking out negative space. During the MRET campaign, we undertook an opinion poll that examined public support for different energy options. The results indicated 95 per cent support for renewables such as wind and solar, 50 per cent for gas and about 21 per cent for coal. The poll was sent to a huge database of politicians around the country. The message: renewable energy is popular, supporting its increased use will also be popular, harming the future of renewable energy will be unpopular and potentially politically damaging.

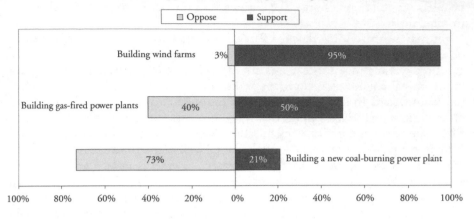

Level of Support for electricity generating options

Source: ARG (2003)

Figure 5.7 *Blocking out negative space with poll
showing public support for renewables*

The following is an example of an attempt to secure positive space on the issue of employment. With a new energy supply on the scene, it might be natural to assume that jobs in other sectors might be lost. To address this assumption, we commissioned a report that compared job creation in wind with job creation in coal; this proved that on a kWh basis wind farms created twice as many Australian permanent jobs and six times more manufacturing and installation jobs! This example shows how we can re-open closed space, which can be difficult because it means tackling areas for which a standard has already been set.

On the issue of cost, many lobbyists have worked very hard to attempt to show that the costs of renewables make them unviable. In fact, significant information suggests this is far from the case. So we prepared briefings and lobby tours to set the record straight based on the reputable economic modelling available.

A counter-attack from the coal and aluminium industry associations followed, which attempted to disprove the original research. Their critique was distributed widely to politicians – but interestingly not in public. So it took a long time to find out what was going on. So, once again, we had to show that their critique was unfounded and commission more work to prove our point. This illustrates how both the proponents and the opponents of the debated renewable energy legislation knew the political space on economic cost was absolutely critical for a positive decision. Here the financial resources of the anti-renewable fossil fuel/aluminium lobby, used to pay for big brand economics reports (of questionable factual accuracy), was almost impossible to match.

Tactics to engage the public

It can be useful to divide the public into three groups: those who support environmental initiatives and clean energy, those who do not or who have higher priorities, and what can be perhaps the biggest group of all – those who lack an opinion on the topic. All three are important and must be engaged.

Some of the tactical elements used to engage the public in the MRET campaign might prove useful here. First, we set out a clear and intelligible target, a 10 per cent increase in new renewable energy production (above a 1997 baseline) adapted from previous targets by industry, NGOs and political parties. The target specifically for wind was bitten off at 5000MW, half the total of the proposed 10 per cent target increase. We made this target more intelligible in public and media materials by stating the target in terms of numbers of homes.

We also agreed a common platform across many allies, as it was crucial to present a unanimous front in the MRET campaign. This was achieved by a series of negotiations which led to an agreed-upon platform ultimately endorsed by more than 50 organizations, including doctors' groups, unions, business and finance groups, students, environmental groups and industry.

> *We the undersigned agree that:*
>
> *I. Global climate change is recognised by world leaders as the greatest global challenge facing humankind in the 21st century. Climate change carries with it serious economic, social, environmental and health risks for all people and nations. In Australia, rising temperatures and extreme weather events such as droughts, floods and severe storms will increase the incidence of bush-fires, cause significant loss of agricultural production, jeopardize natural assets such as the Great Barrier Reef and increase the spread of infectious diseases. Such climate impacts are already leading to increased risks to property and an increased burden on our economy and social fabric.*
>
> *II. Solutions to climate change are available now. One fundamental solution is the move to clean renewable energy sources such as wind, solar and sustainable biomass. Renewable energy resources are proven, low cost and are one of the fastest growing industries worldwide. Australia is blessed with an abundance of renewable energy and is well positioned to be a major renewable technology manufacturing and export centre, creating thousands of high-quality local jobs. Many nations have already adopted policy measures to accelerate the introduction of renewable energy.*
>
> *III. The Federal Mandatory Renewable Energy Target (MRET) review is an opportunity to accelerate the necessary move to clean, renewable energy sources. A 10 per cent MRET in 2010 would be consistent with European and US targets. It would start Australia on the necessary trajectory for significant greenhouse gas reductions. It would also create over 14,000 jobs and establish Australia as an Asia-Pacific hub for clean energy products and services. In the interests of current and future generations, we call on the Australian government to seize this opportunity and increase the MRET to 10 per cent.*

Source: Transition Institute

Figure 5.8 *Banner of thewind.info ezine used in the MRET campaign and subsequently adopted as the AusWEA monthly newsletter*

Finally we made a call to arms. If you alert people to a battle, you had better have something for them to do when they decide to join in! In this case our request was that people actually prepare official submissions to the MRET review process via a joint campaign website prepared on behalf of all the signatory 10×10 coalition organizations.

While there were 120 substantive submissions to the review process, this figure ballooned out to over 5000 when contributions from the public campaigns were counted. Indeed, the secretariat was obliged to take on more staff to deal with the large number of submissions.

Tactics to address sources of concern

It is very important to cover all bases before embarking on any type of public campaign – a sort of risk management strategy. In the case of the wind industry, there is a need to address issues around birds, noise, landscape, tourism and property prices.

The critical aspect here was to break a cycle of claim and counter-claim by referencing as much information as was available on an issue, creating fact sheets, and updating these fact sheets on a regular basis. The aim was to create a definitive source of information which was as unbiased as possible despite the authorship.

The tourism issue is highly subjective and socially driven, and was thus far harder to close off than other wind farm issues. To learn how to address these issues we sought discussions with a nature and heritage conservation group, the National Trust, as previously described in Box 3.5. At the time the National Trust branch in the Australian state of Victoria was pushing for a moratorium on wind power, and this made engaging them more difficult. Nevertheless, any conservation group must be aware of climate change impacts and we found sufficient overlap in our respective missions to engage. This led to an initiative to investigate environmental protection through wind power without compromising landscape protection. The joint project to investigate issues was funded by the federal government.

Tactics to end misinformation

Ending speculation and misinformation on particular issues may require more depth than a simple briefing. We dealt with this challenge during the first year of the 10×10 campaign by producing reports on topics such as the cost of increasing the MRET, jobs, cost convergence of wind and coal, climate impacts, grid and electricity system issues. They were developed for target audiences including those concerned with employment creation, inward investment and risks to that investment, and for farmers and farming communities. Reports like these can fill critical information gaps; in this particular campaign, the reports used became the basis of revised government agency pricing of wind power and were even cited in a Pentagon-commissioned study.

Polling as a tactic

If renewable energy opponents are vocal and are prominent in the media, it may be necessary to test their claims and assert any counter-claims. In Australia, as has been the case in the UK and the US, a small but very vocal group of anti-wind organizers were capturing media attention and appearing to speak of behalf of entire communities. How widespread were the views they put forth? Addressing the type of issue is critical to maintaining political support in the face of sustained criticism. To this end, we commissioned a 1000-person poll into attitudes on various energy issues surrounding renewable energy. The results were unequivocal, showing 95 per cent support for the installation of wind farms as a way to meet future energy needs.

Direct communication to get around media constraints

An ongoing challenge in any campaign is getting the information to the right people. In fact the use of mainstream print media is a very blunt tool. A more sophisticated tool now available is of course email – although it should not be abused. Obtaining the email address of a person you want to send information to, such as a politician or a union official, is usually straightforward. To use this direct communication tool in the MRET campaign, we created a monthly campaign newsletter, or ezine, extolling the virtues of a renewable energy industry and letting critical people know the levels of activity underway.

An example of how effective this tool became was the size of the online list which reached almost 2000 (including almost every state and federal politician in the country) and the attention paid to the newsletters. In one issue we released a map of Australia showing wind investment in each electorate. Over 400 were downloaded in the first hour.

Schedule (When will we do what?)

Clearly the schedule of a campaign will be defined by the objectives. If the objectives in turn are dependent on external circumstances then these will define the schedule. This is especially true if there are windows of legislative opportunity.

Clearly the schedule will be bounded by the available budget as well. Some elements that may define a legislative schedule include timing of government reviews, parliamentary sitting times and state or national elections.

Clearly significant lead times are involved for some campaign elements discussed above, such as reports and websites. Accessing some media, such as major monthly magazines, can entail a six-month lead time. Scheduling becomes a central part of a functional campaign if it is to operate within its resources.

A key area of weakness in scheduling is the ability to know with certainty how long a campaign will indeed take. For example, the MRET campaign was scheduled to last for six months, but more than 18 months passed before a decision was made on the legislation.

Resources and budgets
(What human, financial and other resources are required?)

Estimating the resources required to run a political campaign is one of the most difficult areas to address for a new industry trying to break into a well-established sector. A colleague of mine once pointed out that one person, a computer and a telephone can be a devastating political weapon. However, I am no longer convinced that an industry can get away with this type of one-man-band approach. A critical aspect of having to work to promote change (as in a solutions-based campaign) rather than merely block change puts the onus on the proponents to prove their case, and this is very resource intensive.

Money or people?

A campaign of any sort can be run internally by retaining a good spokesperson from the ranks of industry, an able press officer to get the message out and suitable support. Alternatively, these roles can be outsourced to a PR company which performs these tasks professionally. However, both routes cost money, and the better the standard of the people, the more expensive they will generally be. Industry promotion budgets of any kind often run into the millions of dollars in the industrialized countries. Although they may be less costly in an absolute sense in other countries, the relative costs of the local renewable industry's resources and the cost of services will nonetheless likely stay the same.

The environmental community is not well known for its liquidity, but it will often have significant and skilled human resources for press work, submissions and briefing writing, and lobbying. So even if they lack significant disposable funds, environmental NGOs are likely to have the human infrastructure to take on political campaigning.

The new renewable industries, on the other hand, may have very little to contribute initially. Is there in fact an industry association? Is there a volunteer board? Is there a press officer? Is there a budget for travel and printing and commissioning reports?

This is where a crucial step must come. The industry must pool short-term resources if it wants a short, hard, push campaign to secure positive legislation.

This will far surpass the resources normally committed to an industry association's running work – even if that association has a full-time employee. A prudent businessperson might baulk at the idea of laying out hard-earned money with no guaranteed result.

However, as we couched the issue with Australian renewable energy corporates: 'If we succeed, the market created will make the campaign investment look like petty cash. If we fail to secure or significantly expand the legislation, your ability to grow your business be limited, and spending that money elsewhere won't change that reality. Thus this campaign is a low-risk investment.'

We must also remember that many renewable energy manufacturers are actively looking for new markets and will have discretionary market development budgets that may be suitable places to look for funding.

In addition to financial resources, there are other resources that an industry campaign can draw upon. Because the industry's members may have a very small asset base to work from, these assets must be exploited to the full. The people who comprise the industry, and especially the captains of that industry, are absolutely critical in lobbying. It is they, their businesses and their employees that are on the line in the policy debate.

Industry leaders can add weight to campaign messages by telling a politician about their factory, the people they employ, their aspirations and projections for the future, or their experience from overseas. The industry can also maximize the effects of the campaign if their staff engage in debates that occur in the media and government – by letter writing, media interviews, company-based lobbying, as well as adding in human resources at peak times of campaign activity. All of this makes it generally useful to have good lines of communication between the campaign team and industry, to make the most of all activities in the sector.

Personally I prefer structuring such campaigns as semi-autonomous structures, under the umbrella of an industry organization. This allows the campaign team to get on with the agreed projects under the campaign platform in the real-time cut and thrust of campaigning. It also means that once campaign sponsors have agreed the project parameters and deliverables, they can leave the process to run independently and not have it subjected to influence by their competitors in the industry. These campaign activities can be treated as special projects with their own ring-fenced budgets, administration and reporting, separate to the regular budgets of the industry association.

Measurement criteria (How are we doing?)

A political campaign can be a very difficult beast to measure; and winning or losing may not be the best milestones of impact or success. Public relations people often note that it is easy to be sidetracked chasing leads and reacting to events, to the exclusion of actually working through the agreed strategy and rolling out the tactics. I therefore suggest two types of measurement: first, the actual delivery of the project and second, the impact of that project.

Ensuring delivery of the project

The first step of measurement should cover basic project management. For example, have materials been produced as planned? Have the right people received this information? How many politicians have been briefed and so on? Before we can determine whether a strategy has worked, we have to know whether that strategy has been properly implemented.

Measuring the impact of the project

Measuring a project's impact entails the use of external indicators to help gauge the progress of the campaign with its key messages and target audiences. The obvious (albeit simplistic) indicator in any communication strategy is 'column inches', the amount of coverage in print media. We need to bear in mind, however, that we are trying to affect our target groups' opinions.

A more suitable set of external indicators needs to reflect and be weighted towards what those target groups take from the work done. Useful indicators might include the amount of general and/or specialist media (print, radio and TV) garnered. We can also look at the success of public engagement tools (petitions, websites, letters and postcards), or count the numbers of people receiving direct communication via newsletters. We can quantify the public and private responses or stated support by target audiences. We can furthermore examine the statements or outcomes of government processes, reviews, committees and working groups. And finally we can carry out polling of public or target groups; here, however, we must ensure questions are constructed with care to avoid push-polling, or biasing, of answers.

For example, the outcome of the AusWEA MRET campaign can be judged in part by the significant media attention gained, which generally raised the profile of renewable energy and wind power in particular. As described above, the campaign also prompted a considerable public response, including thousands of online submissions to the MRET review.

While the campaign failed to secure an increase in the target, the MRET was nonetheless retained, one of the major campaign goals. A measure of the campaign's influence on decision-makers may be interpreted in the long delay from the completion of the MRET review to the announcement of a decision. In fact, leaked minutes of the secret May 2004 meeting described in Box 5.1 indicate the government was concerned about the strength of public and industry sentiment towards renewable energy. However, the refusal to raise the target indicates the stronger political positioning of the fossil fuel and energy-intensive industry lobbies in Australia, groups which clearly had stronger access to political decision-makers in the government.

Through reports and proactive and reactive media the campaign also confronted misinformation about wind power projects, again gaining considerable media space and providing a more factual and balanced picture of wind development to the public. Higher-level initiatives, like the landscape project were important initiatives that addressed stakeholder concerns and hedged against future potential conflict.

> ### **Box 5.10** *Two high-profile environmentalists debate wind power*
>
> The following newswire excerpt of an AAP news story illustrates how anti-wind power activists can capture media attention:
>
> > *Two prominent international environmentalists are butting heads over whether wind farms should be set up across Australia. Renowned British botanist David Bellamy has become a strident anti-wind farm activist, denouncing them as pointless, expensive, ugly and dangerous to birds... He has called the advocates of wind-generated power liars and in an Australian newspaper recently described wind farms as 'weapons of mass destruction'.*
> >
> > *Now Canadian geneticist, broadcaster and environmental guru David Suzuki has attacked Bellamy's stance, saying it makes no sense. 'To call wind turbines weapons of mass destruction is unscientific, irresponsible and simply wrong,' he said in a statement. 'Wind farms are about the most environmentally benign energy sources we have – they literally create electricity from fresh air.'*
> >
> > *Suzuki said millions of Australian homes could run on wind-generated electricity by the end of the decade. 'This significant contribution is desperately needed because 84 per cent of Australia's electricity comes from burning coal,' he said. 'Greenhouse pollution from power generation is massive – about one third of Australia's total.'*
> >
> > *The spat between Suzuki and Bellamy over Australian power generation is part of a wider international argument between green groups over the merits of wind farms...*
> >
> > *Other environmentalists have labelled the anti-wind farm lobby NIMBYs (not in my backyard) more worried about local property values than the looming threat of climate change.*
>
> *Source:* AAP (2004)

Can a campaign be counter-productive?

What happens when an issue appears to backfire? For example, what if critics of a certain project or industry use the increased profile of a renewable industry as a hook to place opposing stories and letters? Is more such negative media a sign that the campaign is not working or worse still, that it is being counter-productive?

There is no single answer to this question as it depends on the issues and the industry. In some cases the industry may determine that it cannot expect significant public support by this type of public exchange, and that its interests are therefore best served by keeping issues out of the media. This sort of outcome might be seen in the nuclear industry where the less news about a project, the better its chances.

However, the renewable industry can expect significant public support since it is meeting a crucial environmental need. Furthermore, this industry must actively develop and nurture that public goodwill and understanding in order to facilitate legislative support. Thus renewable proponents should not shirk public debate on renewable energy even if some is negative. That said, the industry must ensure it has the resources and materials in place to rise to the occasion when the debate does get underway.

Although the debate described in Box 5.10 could be said to drag out criticism of wind power, it nonetheless places wind power's impacts within the climate change context, and takes wind farms beyond a mere industrial intrusion on the countryside.

Dynamics (How do we keep abreast of changes?)

Finally we must expect that external events will change the terrain upon which we have mapped our course. We might find the situations for the different target groups have evolved differently than we expected or planned for. Indeed, the more successful the campaign, the more likely there is to be a strong reaction from the proponents and opponents. Thus to ensure wise use of resources, the campaign strategy cycle must be periodically reviewed and updated.

This process may be more difficult than it appears. Once underway, campaigns quickly expand to occupy all available time and resources. However, a strong strategy and implementation plan can help maintain focus on the original plan, keep the necessary discipline to stick to core business, and help gauge which opportunities to seize or let pass. To achieve the dynamic balance of a campaign one must not let the heat of the moment swing the campaign around so much that strategy is lost. Yet one must also avoid campaigns that are too inflexible to evolve with events.

Box 5.11 *SUNTEC: Using fiction to underscore fact*

An EU study released in the mid-1990s by a team led by BP Solar indicated that the challenges facing PV's progression to a lower-cost technology had more to do with economies of scale than with technology constraints. This differed substantially from conventional wisdom that said what solar needed was a technological breakthrough.

This knowledge had very significant policy ramifications since it indicated government support should focus on market development rather than R&D. Clearly this would also have repercussions for existing and prospective companies and investors in the energy and electronics sectors.

Given these identified target groups, the next question was how to relay this information to them in a credible and intelligible form. First, Greenpeace Netherlands commissioned KPMG to use the results of the *MUSIC FM* study to calculate whether enlarged PV markets would lead to costs comparable to the existing delivered cost of electricity in the EU.[2]

The results were compelling. Rooftop solar panels using existing technology would be competitive against conventional generation if manufacturing levels were raised to 500 megawatts peak (MW_p) per year, enough product for about 200,000 houses per year. The authors based the analysis on conditions for northern Europe.

(noted for its lack of sunshine). They also noted that this number did not exceed roofs built or replaced in any given year in a small country like the Netherlands KPMG's point was that from a business angle, solar PV could be considered to be like mobile phone technology; starting with an expensive product and small market, but with the basic technology in place the market would grow and costs would decrease in tandem.

The report was released internationally and received coverage in the business and finance media. A key aspect in its acceptance was that the news came from KPMG, a credible name in the business sector – one of their own so to speak. This report consequently reached the grassroots of the finance sector, the investors in the types of energy companies that might be in a position to act – the ultimate decision-makers.

At the report's launch a London-based marketing company called Cosmonaut were commissioned to help communicate the concept. The brief: 'There is a new energy company that is about to roll out competitive rooftop solar power. Design the pitch to investors and the first advertising campaign.' Some of the results are shown in Figure 5.9.

Imagine if solar power was so simple and cheap that everybody would have it on their roof. Major business analysts have concluded that if solar were mass produced this would become a reality. Greenpeace is campaigning for BP Amoco to make the required investment. Climate change is real and so are the solutions, its time for our energy companies to get real too. **switch on to solar**

Source: Greenpeace (1999)

Figure 5.9 *Promotional materials from the fictional company Suntec (given to BP shareholders to illustrate what BP might offer if it were to expand its PV operations)*

Conclusions

At the outset of this chapter we recognized that working in politics is quite different from working in other spheres, and in order to understand its rationale we must understand the motivations and pressures upon politicians, how they are influenced, and how we can affect those processes. In this chapter we used questions that would be used in any political or marketing campaign as a guide to developing a strategy for promoting renewable energy legislation, and illustrated these with practical examples from a campaign in Australia (which had its own successes and failures).

In some ways this is where the theory part of the book ends as we instead look at practical experience from around the world in the words of experts who have lived and breathed the process of policy evolution in diverse circumstances.

Notes

1   The ECU (european currency unit) was conceived in 1979 as a common European currency and in retrospect can be considered the predecessor of the Euro.
2   Multi-megawatt up-scaling of silicon and thin-film solar cell and module manufacturing (APAS RENA CT94 0008). Coordinators Dr T. M. Bruton and Dr J. M. Woodcock, BP Solar.

References

AAP (2004) 'Green gurus row over wind farms', *Australia*, 11 May
ARG (2003) *National Renewable Energy Quantitative Research*, Australian Research Group report on polling for the Australian Wind Energy Association, September
Frew, W. (2005) 'Full steam ahead a threat to rivers', *Sydney Morning Herald*, 4 June
Greenpeace (1999) *Suntec Investors Brochure*, Greenpeace International
Republic (2003) *Perceptions Audit of Industry and Government Attitudes Towards Wind Energy in Australia*, text slides presented by Republic Consulting, AusWIND conference
Sydney Morning Herald (2005) 'Full steam ahead a threat to rivers: WWF', article by Wendy Frew, June 4

6

A Harsh Environment: The Non-Fossil Fuel Obligation and the UK Renewables Industry

Gordon Edge

Introduction: The Birth of Renewables Policy in the UK

The waiting resource

The UK has one of the world's best and broadest renewable resources. It has 40 per cent of the entire wind resource of the EU 15 (before the recent expansion), massive opportunities to exploit wave power, potential for the use of forestry and other biomass, and a similar solar regime to other northern European countries. Despite these comparative riches and early programmes to exploit them by successive governments, Britain has lagged behind most of its EU partners in establishing a renewable energy industry, especially in the area of manufacturing. If policies were established and financial resources were made available, how has this discrepancy between intent and delivery come about?

The pressing need to address climate change has not receded, and the resolve of the UK to meet its climate commitments is now bringing about the conditions for the country to start catching up. In fact the UK is one of only two EU countries on target to meet its climate change commitments – although it will only do so if its renewables targets are reached. However, the missed opportunities of the 1990s may mean that manufacturers in Denmark, Germany and Spain will be the main industrial beneficiaries of current and future wind energy deployment in the UK. Furthermore, while other renewables are still very much required, they remain in their infancy. There is much to be learned from previous mistakes if they are not to be repeated.

The cultural context

It could be argued that the UK's inability to fully develop its renewable resources is merely one more example of the country's reluctance to exploit its academic brilliance for the benefit of its manufacturing industry. A popular 'truth' in the UK is that innovative ideas are born in the universities across the country or upon the desks of inventors and entrepreneurs, only to lie ignored by the governments and companies needed to support them. However, as we shall see, renewable energy was identified for backing early on in the European development of many energy technologies. The problem in the UK was not that renewables were ignored, but rather the policy support they received did not work as effectively as was intended.

We shall also see that the political climate was central to the type of drivers that renewable energy received. The renewable policies were established during the first stages of power sector deregulation, a fashion that has since swept through the entire industrialized world. Thus an embryonic renewables industry was established during the most radical change to the management and ownership of the power sector since its inception. The primacy of financial efficiency and the use of the market were not confined to the electricity sector, but were being tried in many different areas – even in natural monopolies such as the train service and water utilities. It was therefore little surprise that a market-based model, even if untested, was to be used as the driver for renewable energy.

Finally, the standing and rights of the individual within the legal culture of the UK became highly relevant in respect to development of the renewables industry. The legal power of landowners and residents at a local level proved to be well capable of thwarting national initiatives and national environmental priorities. The failure to recognize this issue and provide a framework for stakeholder engagement and then a forward-planning resolution may have been one of the most significant oversights.

The pressures for policy and the birth of the NFFO

Prior to the privatization of the UK electricity supply industry in 1990, the Central Electricity Generating Board had undertaken research into renewable generating technologies alongside the Department of Energy. This supported a few pioneering companies in the wind sector, as well as innovation in some British universities. The comparatively small seeds of a potentially thriving industry had been sown.

However, the issue of developing renewables was not high on the agenda when the government turned its attention to selling off the UK power industry. The Conservative administration was keen to implement its ideas on competition in the power sector and the government planned to include nuclear generation in the sell-off. However, as the government's plans were analysed by the financial sector, it became clear at a relatively late stage that private finance would not accept the risks and liabilities associated with nuclear power.

The government was forced to retain the nuclear generation business in public hands while it sold off the other generating assets. However, just as the financial analysis had concluded that nuclear generation was a financial liability, so the government had to find a way to cover the cost of this liability. Thus the idea for the Non-Fossil Fuel Obligation (NFFO) was hatched.

In order to ensure that the nuclear fleet could continue to operate and generate the money needed to cover decommissioning costs, the distribution companies would be obliged to take the nuclear output. Since nuclear generation was more expensive than conventional alternatives, forcing the distributors to buy such power would be anti-competitive and therefore undermine the basis of the new market being established. The answer was to cover the above-market costs of nuclear through charging a tax on all fossil fuels – the Fossil Fuel Levy (FFL). This was effectively the world's first carbon tax.

The levy was imposed on fossil fuel-based power, and set by the independent electricity regulator. Through this mechanism, the government made all consumers pay the extra costs for the 'benefits' of nuclear production by applying a broad carbon tax on the rest of the sector. Of course, the scheme ignored the waste and risk problems associated with nuclear power.

As this drama played out in the late 1980s, environmental issues were climbing on the political agenda, as first ozone loss and then climate change were perceived to be pressing global problems. Margaret Thatcher, a trained scientist, was making speeches on the dangers, most notably one to the Royal Society in September 1988,[1] and environmental groups were piling on the pressure. The opportunity was seen within government to extend the use of money made available by NFFO to technologies other than nuclear, that is, renewable energy and thus assuage environmental critics of the government.

At the start, the proportion of NFFO resources made available to renewables was token: over the 1990s, nuclear received £7.8 billion from the Fossil Fuel Levy, while renewables got £400 million. The Fossil Fuel Levy was running at around 10 per cent while it was supporting nuclear liabilities, but dropped to under 1 per cent once only renewables benefited. Nevertheless NFFO proved to be a powerful incentive to a small industry. And by the end of the decade the shoe would be on the other foot, with industries such as wind eclipsing nuclear generation in growth and even price. As the renewable energy technologies expanded, more and more money became available for investment (see Figure 6.1).

The key aspects of the NFFO concept can be identified as follows:

- The UK government would theoretically secure the largest amount of renewable generating capacity for a given cost.
- By giving developers secure long-term contracts, financing would be relatively easy to come by and cheap.
- The price of each technology would be revealed, and through successive competitions, driven down.
- The technology-banding mechanism allotted a predetermined portion of the NFFO pie to each of five renewable technologies. Renewables at different levels of commercial maturity and pricing would not be squeezed out, but would be provided a certain share of the resources.

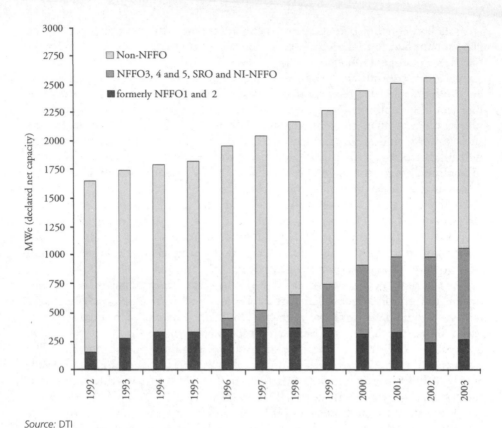

Source: DTI

Figure 6.1 *Renewable generating capacity 1992–2003, including former NFFO contracts (and equivalents in Scotland and Northern Ireland) and capacity outside of NFFO*

The NFFO: Policy and Application Mechanism

Mechanism

To select which renewable projects to support under the NFFO mechanism, the government called for project developers to bid for contracts. Based on the projects' price per amount of energy generated, the cheapest proposals were selected first until the allocation of capacity within each technology band was used up.

Over the lifetime of NFFO there were five rounds of this bidding process. The first bidding round (NFFO1) took place in 1990, with the others following in 1991, 1995, 1997 and 1998. Note the large gaps between rounds, which is something that we will return to later.

The winners were awarded power purchase contracts at the size and price level of their bids. Meanwhile, however, the European Commission had been pursuing a programme of ending national subsidies to the energy sector as part of the path towards a single European power market. The European Commission ruled that the NFFO was a state subsidy to the nuclear industry and had to end in 1998. Thus any developers who won NFFO contracts in the first two renewable tender rounds would gain contracts that only lasted until 1998. The effect was to give only very short periods during which the projects could recoup their investment and this of course led to high bid prices. The Commission later allowed the UK government to extend the scheme for renewables only, and later contracts were awarded for 15 years, with a commensurate drop in bid prices.

The NFFO process

The NFFO process, the outline for which was also followed for corresponding Scottish and Northern Irish Obligations, proceeded as described below.

First the Department of Trade and Industry (DTI; the ministry with main responsibility for energy after the demise of the Department of Energy) would call for bids in an NFFO round. The DTI would indicate how much capacity it would like to see in each of a number of technology bands.

Project developers would then tender proposals for generation capacity in these bands – large wind projects and small wind projects, hydro, landfill gas, waste-to-energy and biomass, including energy crop projects. The third Scottish Renewables Obligation (SRO) round also included wave power. The DTI would review these bids, and take them in order of cost, cheapest first, in each band, up to roughly the amount indicated in the initial announcement of the round. The final totals would not necessarily match the indicative amounts, depending on the amount and quality of bids in each band.

Contracts would then be signed with the successful bidders, which were good only for the site stated in the initial bid. Once this process was completed, renewable generators would sell their output to the Non-Fossil Purchasing Agency (NFPA), an organization owned initially by the 12 regional electricity companies, which were the post-privatization distribution utilities in England and Wales. To cover the difference between the sale price and the contracted cost, the NFPA would sell on the power to the regional electricity company's area in which the project was located at a price set by the regulator, and would receive the funds collected for renewables under the Fossil Fuel Levy.

Contracted capacity

The original NFFO applied only to England and Wales; due to the peculiar nature of the country structures with the UK, separate legislation was required for Scotland and Northern Ireland. This was implemented in 1994, and there have been three rounds of the SRO in 1994, 1997 and 1999, and two rounds of the Northern Ireland Non-Fossil Fuel Obligation (NI-NFFO), in 1994 and 1996. The total number of projects and capacity contracted are shown in Table 6.1.

Table 6.1 *Total numbers and capacity of projects given
NFFO contracts, by NFFO round*

	Number of projects	Capacity (MW DNC²)
England and Wales		
NFFO1 (1990)	75	152.1
NFFO2 (Late 1991)	122	472.2
NFFO3 (1995)	141	626.9
NFFO4 (1997)	195	842.7
NFFO5 (1998)	261	1177.2
Total NFFO	*794*	*3271.1*
Scotland		
SRO1 (1994)	30	76.4
SRO2 (1997)	26	114.1
SRO3 (1999)	53	145.4
Total SRO	*109*	*335.9*
Northern Ireland		
NI-NFFO1 (1994)	20	15.6
NI-NFFO2 (1996)	10	16.3
Total NI-NFFO	*30*	*31.9*
Total	**933**	**3638.9**

Source: DTI

NFFO in Action

Installed capacity

In practice there was a stark difference between the NFFO-contracted capacity and the actual delivery of installed capacity. By the end of 2003 only 30 per cent of the contracted capacity on a declared net capacity (DNC) basis had been built. Had all of the projects awarded contracts been built, the UK would have 1154MW (DNC) of wind (nearly 2700MW of nameplate capacity) and hence been a considerable market for this technology. By 30 June 2001 only 164MW (DNC; about 380MW nameplate) of this capacity had been commissioned. Technologies other than wind have generally been more successful in getting through to completion, although waste projects have also had severe problems (see Table 6.2). This throws up an important area of policy weakness that we shall explore in due course herein.

Table 6.2 *Total numbers and capacity of projects given NFFO, NI-NFFO and SRO contracts and completed, by technology*

Technology	Contracted projects		Commissioned projects (as at 31 March 2004)	
	Number	Capacity (MW DNC)	Number	Capacity (MW DNC)
Biomass	32	256.0	9	106.5
Hydro	146	95.4	68	47.4
Landfill gas	329	699.7	226	474.8
Municipal and industrial waste	90	1398.2	20	235.5
Sewage gas	31	33.9	24	25.0
Wave	3	2.0	1	0.2
Wind	302	1153.7	93	219.8
Total	933	3638.9	441	1109.2

Source: DTI

Pricing

One of the most successful elements of the scheme is that it did allow some insights into the real pricing of renewables and also did succeed in its stated aim of economic efficiency, providing a pressure to keep bid prices as low as possible. In Appendix B are tables which illustrate how bid prices did drop in successive NFFO rounds – although much of the drop from NFFO2 to NFFO3 can be attributed to the increased length of contracts offered after 1994.

Having said this, we need to note that NFFO itself was not necessarily causing these prices to decrease; rather cost reductions in successive NFFO rounds had a lot to do with economies of scale, increasing industry experience and technology advances being driven by the larger markets. In effect, the UK government ensured that the NFFO scheme reaped the benefit of the industry development money spent in Germany and Denmark, which pursued less competitive and therefore lower risk market drivers that attracted industry development and investment (see Figure 6.2).

The incubation of new technologies

One of the notable elements of NFFO that differed from the Renewable Portfolio Standards and similar competitive market-based schemes was the inclusion of technology banding. This was very deliberately intended to ensure that the resources produced targeted industry development and did not lead to the least-cost technologies taking all of the resources.

An interesting example of this was the identification of wave power for development. Although it can be argued that many of the other technologies had been

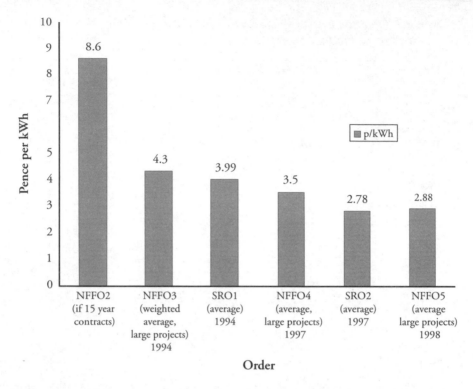

Source: DTI

Figure 6.2 *NFFO benefited from industry development money spent in Germany and Denmark which, for example, resulted in falling prices for wind power*

developed to a greater or lesser extent with other countries, wave energy was one source that, despite its significant potential as an energy to be harnessed, has languished beyond the interest of governments throughout Europe. By bringing wave into the rounds, the government hauled the wave industry onto the fast track of technology and industry development.

Although we are now used to seeing feed-in law-type systems using technology definitions to set prices for power, the NFFO family of schemes is a rare example of a competitive market-based system using such specificity. The fashion for competitive systems is now trending instead to quite a different approach of technology neutrality where price alone dictates the technology mix.

Manufacturing capacity

The policy of awarding contracts to the lowest-price bidder had a negative effect on the ability of UK industry to develop its manufacturing capabilities in these new technologies. With developers forced to choose the technology that would deliver the lowest price in order to secure NFFO contracts in competition with other projects, there was no slack in the system that would have allowed some to choose higher-cost local manufacturers or in-house technology. There were also no guarantees about a future stream of possible developments. It was a high-risk environment for all concerned, with short-term time horizons.

While UK companies were very early off the mark in the technical development cycle, even compared to the Danish wind turbine makers, the extreme price pressure on companies meant that manufacturing ambitions were quickly abandoned or sold on. NFFO was thus a lost opportunity to nurture an indigenous renewable technology industry.

Service industries

The counter-effect of a hostile business environment is that those companies that successfully learned to operate under NFFO have become very successful at competing in new markets. Under NFFO, the developers and consultants honed their skills and sharpened their ability to reduce project prices given a set of hardware costs that were similar for all developers. Furthermore, having built their businesses in the liberalized UK environment, they perhaps have some relative advantage over their German or Spanish competitors in being able to successfully adapt to liberalized or indeed liberalizing power markets.

These companies are now taking their capabilities abroad. Renewable Energy Systems, a subsidiary of the McAlpine construction company, has been involved in large projects in the US, and is aiming to be a big player in the French and Australian markets. National Wind Power (now 'npower renewables') has also had success in the US, while consultants Garrad Hassan are highly regarded around the world.

The NFFO experience also resulted in some expertise building in the finance profession. Practitioners became familiar with renewable technologies while bringing to bear their skills in project finance, a form of funding which was facilitated by the long-term NFFO contracts. Different financing structures are needed for NFFO's replacement, the Renewable Obligation (see below), but now bankers and investors are at least familiar with the technologies.

NFFO Policy Analysis: What Could Have Been Done Better?

The NFFO experience highlights the dangers of creating a renewable energy driver without preparing an integrated policy framework to deliver it. Here we consider some of the issues related to integration or 'joined-up policy-making'.

Industry and employment requirements

In common with other European governments, the UK was coming out of a period of very high unemployment when NFFO was being set up. Crucial vote-winning issues were new industries and new jobs. The establishment of a UK technical and manufacturing lead in these new industries and technologies was much discussed, but failed to be expressed explicitly in the policy make-up of NFFO. This may have been a function of the then government's faith in free market economics.

A provision for manufacturing or local content may or may not have been appropriate for inclusion in the NFFO mechanism itself, although EU state aid rules would likely have precluded an explicit provision. However, since it appeared nowhere in the supporting policy frameworks, the ability to ensure the delivery of manufacturing and employment was left beyond the control of government – a significant omission given that the government held the purse strings. This omission stands in stark contrast to the policy of various Spanish regions, where planning support for renewables was successfully hard-wired to the delivery of local content, manufacture and therefore employment.

Integrated planning

The reasons for the difficulty in translating contracts into capacity are complex, but primarily concern the relationship between NFFO and the planning process – or rather the lack of a relationship. Gaining an NFFO contract was only the first step in building a project – it had to be approved by the local authority before construction could begin. With the renewable element of NFFO added as an afterthought to the privatization legislation, little attention was given to the potential problems that developers might have in getting such planning permission. The problems that were revealed by NFFO were also exacerbated by its bidding structure.

Due to the confidentiality that was required in the bidding process, developers were not able to inform local communities of their plans for a project – something that would be considered a suicidal act by current community consultation standards. Often the first inkling that a wind farm or biomass generator would be built was the award of the NFFO contract. In this situation, it was difficult to build the trust of local authorities and residents who would then decide on planning applications.

In addition, there is some evidence that for wind power, the strong emphasis on least cost gave rise to difficulties in the planning process. The need to minimize generation costs drove developers to the very windiest sites – which tend to be areas of high landscape value and hence areas where wind proposals would be more prone to objection and possibly rejected. While some developers dispute that this is a real problem, saying that planning is difficult for any site, it is certainly true that the NFFO bidding process left very little financial room to manoeuvre. This limited the ability to accommodate the concerns of local communities, which did not make gaining planning permission any easier. There was also the provision that a project could not be re-located, again further limiting options for successful delivery.

While the structure of NFFO made life difficult for developers, more capacity would have been built if there had been stronger guidance from central government to local planning authorities. In the absence of this steerage, councils were often swayed by small, but vociferous minorities who objected violently to virtually any application. It was the organization Country Guardian, which has as its most prominent member Margaret Thatcher's former press secretary, Bernard Ingham, who crusaded against wind power. Ingham is also a member of Supporters of Nuclear Energy, fuelling suspicion that the nuclear industry was behind the attacks. A string of planning decisions followed which turned down applications for wind farms on the grounds that the landscape damage that would result was not justified by the small amount of power that would be generated. This was inconsistent with stated government policy, but without this policy being translated into direction in the form of a Planning Policy Guidance note, local officials were free to ignore it.

This lack of forethought on the part of central government to the issue of planning contrasts with countries such as Germany and Denmark, which early on marked out which areas would and which would not be suitable for development. Developers could hence move forward with projects in the confidence that permission would not be refused, within reason.

The model of support in these countries, a simple fixed tariff, with its low risk profile, also enabled many wind turbines to be cooperatively owned by local residents, thus building a constituency of support that facilitated the granting of planning permission. The complex NFFO bidding process was only really accessible to professional developers and financiers, so local communities in the UK were effectively shut out – hardly a situation likely to breed trust and mutual support.

Degree of competition

The emphasis on the lowest bidder created a highly competitive system, encouraging developers to be optimistic in the extent they bid down prices, and it did little to filter out those who had been over-optimistic. With everyone learning by doing, some mistakes were inevitable, especially under the pressure of the competitive process. Given that winning bidders had a grace period of five years before the contract offer expired, some may have bet on equipment costs falling; if the costs did not fall enough, then their projects were not viable. Thus the combination of strong competition with no penalty for non-delivery left precious NFFO resources unspent and locked out less competitive tenders that could have been realized.

Continuity and stability

Also detrimental to the development of a thriving renewable business sector in the UK was the stop–go nature of NFFO. Bidding rounds were irregularly spaced, with relatively little indication of when the next round might occur. It is notable that a review undertaken after the second round left a three-year gap between tenders being invited, during which time many of the fledgling businesses collapsed through lack of business, or left the industry due to dissatisfaction with the process.

When a tender was announced, the risk was all left to the developers. The developers would be thrown into a frenzy of activity, all the while having to calculate how much resource to devote to their projects when there was no guarantee they would be successful. Not only did they not know what their competitors would offer, it was not clear how many projects the government would award. In between bidding rounds, companies had again to judge how to deploy their resources in the face of uncertainty, with the timing and technology banding of the next round in the hands of government ministers.

Financing

Another factor that may have affected the relative success of the British and particularly the German renewable industries is the availability of long-term finance at reasonable cost. In Germany, the public bank Kredietanstalt für Wiederaufbau (KfW) provided cheap loans with suitable term lengths. Alongside generous fixed tariffs and a pool of ecologically motivated small investors benefiting from tax breaks, this has been a major factor in making Germany, with a relatively poor wind resource, a world leader in wind power. In the UK, meanwhile, developers had to gain all their finance from commercial sources, banking on the security of the NFFO contract to lower the cost of capital somewhat. What went wrong?

Again we may want to think of this in terms of risk. NFFO was a series of competitive tenders – the price paid was unknown from project to project, introducing uncertainty. The ability to finance wind projects in an aggressive environment was far from guaranteed and so the business process was itself higher risk, as it is with any competitive tender. These rounds were not rolled out on a pre-planned regular basis, but at sizes and periods at the whim of government – adding more risk. And finally the environment of undelivered projects, planning failure and so forth further increased the financial risk associated with the industry.

By contrast, the massive manufacturing investments required for capital-intensive renewables will always seek lower-risk environments which they found in Denmark, Germany and Spain. In the UK, alongside bid-down contract prices and cautious council officials refusing permits, this financial risk issue contributed to the huge wind resource remaining largely untapped and a manufacturing base undelivered.

Pressure for change

The last NFFO bidding round was in 1998, and the last SRO in 1999. The policy introduced under the Conservatives had been continued by the new Labour government after its landslide election victory in 1997, and there had even been talk of a special NFFO6 round for offshore wind only. But the pressure for change was building due to mounting evidence that the NFFO system had serious problems.

The government embarked on a consultation process to define a replacement for NFFO. This was a long, drawn-out procedure, however, especially under the hand of Energy Minister Helen Lidell, who was generally regarded as giving little priority to renewables. Yet again, developers had to suffer a policy and market vacuum for the majority of the first Labour term of office, which did not give them much encouragement.

The Renewables Obligation

Given the Labour Party's stated commitment to the environment in the run-up to the 1997 general election, one might have expected the new Labour government would have moved swiftly to implement new policies to stimulate the market for renewable energy. However, the process which finally led to the Renewables Obligation (RO) was protracted and characterized by constant consultation.

On entering office, the new energy minister John Battle set out to review policy on renewable energy. Despite assurances that the government wished to move boldly in this area, it was not until March 1999 – nearly two years into Labour's first term – that the review was finally published. It contained the general conclusion that a form of the then relatively untried quota-and-trade system (known in the US as Renewable Portfolio Standards) should be implemented through an obligation on suppliers to buy green power; tradable certificates were to be used as the measure of compliance. This was in keeping with the UK's bias towards market-based solutions.

The general form of what was to become the RO was consulted on before enabling powers were introduced into the Utilities Act of 2000. In order to make the RO operational, however, secondary legislation in the form of the Renewables Obligation Order had to be drafted and passed by parliament, which required yet more consultation. This order was finally put on the statute books in 2002, and on 1 April that year, the RO came into force.

Revised aims

While NFFO was aimed at developing a number of technologies through what were essentially separate auctions, the RO was intentionally set up to be technology blind, with the cheapest resources exploited first. This was in keeping with NFFO's focus on cost as the prime concern; the mantra from government in the RO development process was that renewable energy policy must result in costs to the consumer that were acceptable, although what constituted 'unacceptable' was never spelt out explicitly.

As with NFFO, quantities were defined, although for the RO the quantity was energy generated and not generating capacity, and the market was expected to discover the appropriate price through competition. The major difference between the two policies was the scale of the ambition: the RO would be putting about £1 billion into the renewable sector by 2010. Key problems with the mechanism threatened these ambitions, however, and ongoing issues with the surrounding context of planning and grid infrastructure continued to make life anything but easy for renewable developers.

The mechanism

The RO is a system which mandates licensed electricity suppliers to supply a set proportion of their sales from renewable generation. This percentage started at 3 per cent in 2002–03, and was initially set to rise to 10.4 per cent in 2010–11 (see Table 6.3). In late 2003, the government responded to criticism that this profile

Table 6.3 *The profile of the Renewables Obligation up to 2010–11*

Period	Estimated sales by licensed suppliers in UK (TWh)	Total Obligation in UK (TWh)	Total Obligation as % of sales (UK)
2001/02	310.9		
2002/03	313.9	9.4	3.0
2003/04	316.2	13.5	4.3
2004/05	318.7	15.6	4.9
2005/06	320.6	17.7	5.5
2006/07	321.4	21.5	6.7
2007/08	322.2	25.4	7.9
2008/09	323.0	29.4	9.1
2009/10	323.8	31.5	9.7
2010/11	324.3	33.6	10.4

Source: DTI

would not provide enough incentive to developers, and proposed to extend the Obligation to rise in annual 1 per cent steps to 15.4 per cent in 2015–16. The Obligation then remains at this level until 2027.

The means by which suppliers achieve compliance with the Obligation is by acquiring Renewable Obligation Certificates (ROCs), tradable instruments that are awarded to qualifying generators in proportion to their output, with one ROC equalling 1 megawatt hour (MWh). Renewable generators therefore have two revenue streams – the plain power output, sold at market prices, and the income from ROC sales.

To ensure compliance, if a supplier is unable or unwilling to acquire the required amount of ROCs, it has to pay a buy-out price, originally set at £30/MWh for the shortfall, with the price rising each year with inflation. For 2004/05, the buy-out price was £31.39. The funds raised as a result of buy-out payments do not disappear into a government bank account, but are recycled in a way that adds pressure to the compliance regime. The buy-out funds are distributed pro-rata to those suppliers in accordance with their level of compliance under an arrangement known as the 'green smear-back'. Due to completion problems that many projects have encountered, there is a shortfall in the number of ROCs available, which means that ROCs will fetch a price higher than £30/MWh. Even though this is higher than the penalty payment, those that are successful in securing ROCs will get a higher proportion of the buy-out proceeds, thus offsetting the extra paid over the buy-out price.

It is interesting to note that the recycling of the buy-out penalty to the compliant retailers creates a secondary driver for compliance. That is, retailers who fail to meet their requirement pay a penalty, but this penalty is then distributed among those retailers that have complied, so a non-compliant company is effectively paying its penalty to its competitors. The other effect of the recycling mechanism is to effectively cap the cost of the RO – which was the government's intention – but then also share that capped pot among those who have actually managed to generate ROC-eligible power. This means that as renewable output gets closer to the obligation level, the lower the ROC price will be, thus providing a disincentive to investors to build too much capacity and threaten the return on the plant that has been built. In addition, were the obligation ever to be met, ROCs in excess of the levels required would be worthless, and there is even the possibility that in this situation a race for the bottom in a buyers' market would result in a price crash for all ROCs.

NFFO–RO bridge

NFFO projects from the first two rounds of bidding have been free to sell their output in the free market since 1998 and participate in the RO system fully. Those from the later rounds are still contracted to the NFPA for their output. For England and Wales, the agency is now auctioning the power bundled with the associated ROCs to the highest bidder. In Scotland the Scottish ROCs are auctioned separately, but the effect is the same. For the first years of the RO, they are the main source of ROCs on the market. As capacity is built specifically for the RO, the importance of the NFFO projects gradually subsides.

With the shortage of ROCs on the market in the first years of the RO, prices have been high and the NFPA has been receiving considerable income from its ROC auctions. This income is much more than is required to pay NFFO-contracted generators the prices they bid, and the NFPA has thus accumulated a significant surplus. When NFFO was set up, however, it was not foreseen that the NFPA would be anything other than a means to channel funds collected under the Fossil Fuel Levy to renewable generators, and there was no legal power for government (or anyone else) to claim and spend this money. After amending a private member's bill in the House of Commons (the Sustainable Energy Act 2003), this power was taken and £60 million was directed to renewable energy spending, while the rest of the pot outstanding and any future surplus disappeared into general government coffers. Thus resources that have been mandated for the support of renewables through the RO have been diverted back to the treasury.

Early RO: What can We Learn so Far?

Market stability

Like all market-based mechanisms for support of renewables, there is a price–risk element to the system. If the market is stable, these risks can be calculated and shared among the market players willing to shoulder them.

However, one of the unforeseen issues with the operation of the RO has affected market confidence and resulted in an unstable trading environment. The problem was revealed when TXU Europe, a leading player in the UK power supply market, went into liquidation before the RO compliance procedure for the year had completed. TXU's customer base was sold to another supplier but the bankrupt company remained the liable party for the RO that those customers' consumption had incurred. The buy-out fund became merely one of a large number of creditors making demands on the liquidated assets of TXU and that left the fund short of money to pay out the expected recycling benefits to ROC holders. Trading stopped immediately, and the ROC market has been barely liquid ever since. With an illiquid market, it is impossible to develop the ROC forward contracts and other financial products which would ensure smooth operation and reduce risks. While the government has concluded that the obvious solution to protecting the buy-out fund – having the Obligation follow the customers – is impossible under insolvency law, it has worked on solutions to mitigate the effect of supplier bankruptcies. It is yet to be seen whether these will provide the market with the confidence it requires.

Competition issues

The key to RO success with respect to new capacity will be whether the pressure of competition acts at the supply side or the demand side. NFFO competition was excessively severe at the supply side as we saw. This pressure is alleviated if the RO target is sufficiently ambitious, driving up prices for ROCs and creating higher margins for renewable generation. It is worth noting here that, since so many projects are rejected at the planning stage, those that are successful must make a good return in order to cover developers' expenses for the unsuccessful ones.

Economic efficiency versus industry building

Since the new RO system only went live on 1 April 2002 it is still too early to tell whether it will succeed over the long term where NFFO failed. What can be said is that it continues to emphasize economic efficiency, in this instance depending on trading of ROCs to minimize cost, rather than competitive bidding. Consequently there is a danger that much of the benefit of the new policy will go to foreign manufacturers nurtured in a less demanding environment.

Indications so far point to a more stable market that the RO has fostered, encouraging wind manufacturers to set up operations in the UK. The trailblazer here has been Vestas, which has a successful assembly plant at Campbeltown on the Mull of Kintyre in Scotland, and mergered with NEG Micon, a leading blade production plant on the Isle of Wight (off England's south coast). Since UK engineering firm FKI acquired Germany's DeWind, it has started manufacturing operations in Loughborough, and REpower has a joint venture with Peter Brotherhood. Danish Bonus has also been seeking out possible sites. In most cases, the profits will be sent abroad, although the UK economy will benefit from jobs and investment, as well as business further down the supply chain.

One of the ironies of having a mechanism which claims to promote economic efficiency and seek least-cost solutions is that it can result in higher-cost finance. The prices achievable under the RO are not known with any certainty and uncertainty only increases the further into the future one looks. The prices are also dependent on the overall activity level in the market (which affects the amount of money to be recycled), and are at risk due to supplier default. With bank financing, the higher the risk attached to an activity, the higher the cost of capital will be, and as the cost structure of many renewables (particularly wind) is heavily capital intensive, this pushes up the cost of the power output.

Technology banding

The use of a single market for all technologies (rather than a system of technology banding) when combined with an emphasis on costs means that the RO primarily benefits the cheaper technologies. This is to the exclusion of more expensive sources that may be less progressed in their industrial development. The RO will therefore inhibit their development relative to other sources, rather than cause an acceleration of a diverse set of commercially viable technologies. This is recognized by the UK government and is an explicit part of the RO policy.

If the UK is therefore to preserve its lead in the emerging technologies of wave, tidal stream and advanced biomass conversion, then supplementary policy instruments are required. There are some already in existence, such as R&D grants in the wave and tidal field and capital grants for the first round of offshore wind projects. Others are being debated, such as 'photocopied ROCs', where an additional fixed premium is paid to generators using the technologies being supported. Whether the capital grants available for offshore wind and biomass will suffice is arguable, although the limited amount available for the former has not stopped a frenzy of activity in this sector. As for solar power, the RO will not be much use in developing this market, given the high cost and distributed nature of this technology; the RO favours larger generators due to its complexity.

The policy's structure also requires the addition of complementary measures that engender new companies rather than favouring the existing bigger players. Currently the RO is also biased towards project development by the large integrated power companies, such as npower and Powergen, since they can extract the maximum value from ROCs; by keeping the entire ROC value chain under one roof they get the whole revenue. Independent developers have to sell the output of their projects to suppliers, and since there are only six big companies retailing electricity, they have significant market power and can squeeze developers. This makes life harder for independent renewable developers, who will be likely to struggle in the new order. The complexity of the RO also militates against the development of community-owned renewables, which has been shown to be an important issue in public acceptance.

Persistent planning obstacles

Despite the introduction of the RO, the problems of planning remain. The national government has been slow to take firm control at a policy level and update guidelines for development and planning of renewable energy projects. After much delay, Planning Policy Statement 22 has finally been published, which should weight the balance more towards renewable generators in England, and the equivalent Technical Advice Note 8 in Wales foresees additional onshore wind in the principality. A system of regional targets has been drawn up in England and Wales. The effect of this will still take some time to filter down to the local authority level, however. In Scotland, meanwhile, the devolved administration has been much more proactive towards renewables. The planning guidance recently put in place in England and Wales has been in place in Scotland for years. Scotland is reaping the rewards of its stance, with more development activity there than in England.

Transmission issues

Another issue that is slowly being resolved is the cost of grid connection and the equitable sharing of the embedded benefits of renewable power. Most renewable generators are relatively small and connected directly to the distribution network, bypassing the high-voltage transmission grid. If their output is sold to customers within the distribution area in which they are situated, the power has a higher value since it saves the distribution company from paying transmission charges and it avoids transmission losses. In some instances it can also avoid the need for costly transmission system upgrades.

Due to the way that the distribution industry is structured, these embedded benefits are difficult to capture for the generators – they generally accrue to the distribution network operators (DNOs), who are unsurprisingly reluctant to relinquish them. The regulation of the DNOs encourages them to maximize the amount of power running through their wires, rather than minimize the overall system cost. This was considered in the 2005 distribution price review, but it will be some time before mindsets are changed.

The UK is also coming up against the issues of transmission network extension to where the resources are found, particularly in the case of wind. Most onshore development is in remote parts of Scotland, and requires significant network reinforcement. Similarly, connecting the large-scale Round Two offshore projects (totalling up to 7200MW) will require large investment in new transmission assets. How these investments are paid for and whether they will be complete in time for renewable energy targets to be met are currently big issues in the UK wind industry.

Grid connection issues

Again, this is a central-government policy issue, requiring strategic changes to the grid infrastructure and its management that will be necessary for the transition to renewables.

Current system rules also allow the DNOs to charge 'deep' connection costs – any reinforcement of the system needed to accommodate a renewable generator is charged to it rather than spread over the distribution customers. This of course makes renewable energy more expensive and therefore threatens market development. Moves are now afoot to change this to 'shallow' connection costs – merely the cost of connecting to the network, with any work needed to reinforce the grid taken on by the DNO and the cost spread over all users of the system. Also, in the 2005 distribution price review, the regulatory model provided an opportunity for change so that the DNOs are motivated to minimize overall costs. This should be of major benefit to distributed generators.

Access and priority

Making the changes to the regulatory model discussed above would require the gas and electricity regulator Ofgem to be significantly more cooperative. Going on past evidence, however, it may not be so. Pessimism about the willingness of Ofgem to give renewables priority in the distribution price review stems from its attitude during the design and introduction of the New Electricity Trading Arrangements (NETA).

The Utilities Act that instituted the RO also provided the legal framework for NETA. This system replaced the Electricity Pool that had been in existence since privatization in 1990. Where the pool had been a centralized bidding system, NETA relies on bilateral contracting between generators and suppliers. Trading of these contracts is allowed up to 'gate closure' (currently one hour ahead of real time), after which the System Operator (SO, National Grid) runs a balancing market, under which it pays for extra generation or reductions in demand as appropriate, the costs of which are recovered through transmission charges. If a player's contract position at gate closure is different from the real outcome (e.g. a generator produces more or less than it is contracted to do), then it is charged by the SO. If a generator generates more than it has said it will, then the extra is bought by the SO at the system buy price – which is much lower than the market price. If a generator produces less, the SO buys on its behalf and passes this on as the system sell price – which is much higher than the market price.

It was clear when Ofgem was designing the NETA system that this structure would penalize generators that were unable to predict their output precisely; apart from the obvious intermittent sources like wind, this also applies to combined heat and power (the output of which is dictated by the heat demand and cannot be controlled easily). While Ofgem claimed that the imposition of imbalance charges on intermittent generators reflected the costs such sources imposed on the system, other analyses have concluded that the actual costs are considerably lower and the charges are penal. When the system was introduced, there was indeed a depressing effect on income for such generators – the combined heat and power sector in particular has never recovered from the blow, while wind now has the RO, which allows it to absorb the cost.

The NETA saga shows that despite being an arm of government, the regulator Ofgem has pursued policies which conflict with government objectives because of its narrow interpretation of its remit to protect the interests of consumers. The requirement for Ofgem to take sustainability into account has been strengthened by the 2004 Energy Act, which has also extended NETA to Scotland through the British Electricity Transmission and Trading Arrangements. Whether these changes will make the situation better has yet to be determined, although there are some key tests coming up over the extension of the transmission grid to remote parts of Scotland and out to sea, where much of the renewable resource is to be found.

Conclusions

The UK is a nation with some of the highest renewable energy resources in Europe and the political support for renewable energy was in place very early on. Yet there has been a persistent discrepancy between the intent for a vibrant British renewable energy industry and its delivery. There are lessons that can be learned to ensure that future opportunities for manufacturing industries in new technologies or new renewable energy markets are not missed.

Carbon and competition

The renewables policy framework in the UK was defined by the confluence of three forces: a carbon tax (initially established to support the nuclear industry), a political agenda of forcing competition into energy and other markets, and rapidly developing renewable energy technologies.

The embryonic renewable energy industry rightfully claimed access to some of the funds from the carbon tax, NFFO, and eventually monopolized the entire scheme. But the price was that it not only had to operate in the world's first fully deregulated power sector, it also had to operate as a competitive renewable energy market.

An Englishman's home is his castle

Another defining characteristic has been the ongoing challenge of stakeholders. An unwitting side effect of the initial policies was that stakeholders became marginalized, yet powerful. They were kept uninformed due to confidentiality clauses, and yet they maintained a central role in local council planning decisions. A small but vocal opposition to wind energy emerged as a result. This more than any other oversight points to the need for integrated, whole-of-government policy-making.

Great aspirations

The NFFO was arguably an industry-development focused policy. It had noble aspirations of targeted and weighted development of emerging industries through technology banding, industry security through long-term contracts, transparent disclosed pricing and finally the delivery of the maximum volume of clean energy for the lowest price by using competitive tenders.

The curse of NFFO, however, was that the competitive pressure came to dominate and undermine almost all of the other goals. It gradually created a market in which all projects were competing so closely they became marginal, and many approved projects failed to be built. It introduced loopholes that allowed speculation, predatory practice and left allocated resources unspent. And the fundamental basis of a series of discretionary, indeterminate and unscheduled competitive tenders introduced a boom-and-bust cycle into the market that undermined the long-term stability and outlook which is the bedrock of manufacturing investment. So the political vision was there, but perhaps the will and commitment were not as strong as they could have been.

Desert flowers

A harsh environment encourages hardy survivors. Those companies that did manage to stay in business through NFFO had to hone their skills and competitiveness and have since been able to compete effectively in the growing world renewable energy markets. This has been especially true in the finance, service and project development sectors.

Switching from carbon tax to Renewable Obligation

A fundamental shift in policy occurred when the NFFO pool of money allocated by the government was scrapped in favour of a scheme that put the full onus on the electricity utilities. By requiring retailers to buy Renewable Obligation Certificates, a new market for renewable energy was created.

Although this was a fundamental shift in responsibility, it maintained a scheme based on competition and low-cost impact to end consumers. Its advantages were that it removed many of the loopholes of non-delivery and allowed the market to find solutions to many of the financial or contractual obstacles facing renewable energy developments.

An unfortunate consequential loss was technology banding, which in many ways was the one aspect of the NFFO that was delivering industry development for the smaller and less developed industries such as wave power. Under the RO it is the survival of the fittest and most cost effective.

Joined-up policy-making

Perhaps the most important lesson to be learned from the renewable energy policy evolution in the UK is the need for a framework of policies that address the industry as a whole – rather than just its financial drivers alone. The issues of planning, transmission, grid connection and dispatch are all fundamental to successful renewable energy markets, and if left unaddressed can entirely thwart the best-intentioned drivers.

The British test tube

In many ways, the UK was an early experimenter in using competitive markets to further renewable energy industries and markets. It has not been an easy path for the industry participants. However, at the time of going to press, there have been more megawatts of wind approved for construction in the past year than built in the previous 13 years (770MW). Indeed there are 2.5 gigawatts (GW) of wind approved for construction worth 2.5 per cent of the UK's electricity demand. So clearly there have been lessons learned and progress made.

Perhaps the greatest shame has been the failure to secure more manufacturing industry in the UK. Clearly this has gone to those countries where there has been more financial certainty. Thus if there is one overriding conclusion, it is that the UK policies have largely done what they defined to do in terms of producing the most energy for the lowest cost, but they did not do what they would have assumed they would do, which was fill factories with workers. In the long run, we may ask, was the cost of a short-term saving the loss of a long-term income?

Notes

1 As Mrs Thatcher said in the speech: 'The protection of the environment and the balance of nature are one of the greatest challenges of the late 20th century.' She went on to proclaim: 'For generations we have assumed that the efforts of mankind would leave the fundamental equilibrium of the world's systems and atmosphere stable. But we have unwittingly begun a massive experiment with the system of this planet itself.' And in perhaps the most famous quote from the speech: 'We do not have a freehold on the Earth, only a full repairing lease.'

2 DNC: Declared net capacity is the nameplate capacity of a project multiplied by a capacity factor, set for each resource type. DNC can be regarded as a measure of the firm capacity of intermittent sources, compared to nuclear availability. The capacity factors used are 0.43 for wind, 0.17 for solar, 0.33 for tidal/wave, 0.55 for small hydro and 1 for all others.

7
Renewable Policy Lessons from the US: The Need for Consistent and Stable Policies

Randall Swisher and Kevin Porter

Introduction

This chapter examines the development of wind energy markets in the US and in Europe, and the importance of policy as a driver. In summary, the wind energy industry in the US has seen two periods of growth from two types of policies. However, the policies have not been stable and consequently the wind industry in the US has seen boom–bust cycles of extensive wind energy development followed by periods of little or no development. In contrast, renewable energy policy has been more constant in Europe where wind energy development has been steadier. Europe's is a much more vigorous wind industry as a result, most notably in the area of manufacturing.

The lesson to be drawn is not so much an endorsement of specific European policies; in fact, the most successful European policies would not be politically supported in the US and are not readily transferable given differences in the respective electric industries. The lesson is more that policy support must be consistent, predictable and long term if renewable energy is going to make a significant contribution to the US electric power sector. Such stable policy, leading to a constantly growing market, is essential for any country that hopes to build a significant manufacturing capability in this rapidly growing industry.

The US PURPA Years: The First Flowering

The modern wind industry got its start in the US in 1978, when Congress passed the Public Utility Regulatory Policies Act (PURPA) as part of the National Energy Act. PURPA was a foundation for the independent power movement and, largely due to progressive regulators in California, played a seminal role in helping to facilitate a commercial market for renewable energy technologies such as wind, solar thermal, biomass and geothermal power.

Access, pricing and incentives

PURPA had two components that were critical for renewables: access and pricing.

Regarding access, the law required electric utilities to interconnect to 'small power producers', which included independent generators relying upon either renewable sources or co-generation.

As for standardized pricing, the rules developed for the new law required utilities to purchase power from the independents at what was termed the 'avoided cost' of the utility – the cost the utility would face if it were to build the new generation plant itself or purchase it on the market from another source.

In addition, the National Energy Act included two other provisions that were important fiscal incentives to renewables. The first provision was investment tax credits of 40 per cent for residential wind and solar energy systems, and 10 per cent for business investments in wind, solar and geothermal energy systems, later increased to 15 per cent in 1980.[1] These tax credits could be taken on top of a 10 per cent general investment tax credit. The second National Energy Act provision was accelerated depreciation recovery of five years for investments in wind, solar and geothermal energy technologies.

The role of the states

Implementation of PURPA in the early 1980s led to the 'first flowering' of renewable energy in the US. In issuing implementing regulations in 1980 (after surviving a court challenge), the Federal Energy Regulatory Commission largely delegated the setting of avoided costs to the states. California in turn ordered electric utilities to enter into contracts at standard terms and conditions, that is standard offer contracts. Four standard offer contracts were available; the one most significant to renewable energy generators such as wind was the Interim Standard Offer #4 contract, otherwise known as ISO4. The ISO4 contract was a 30-year contract with a capacity payment, with the first 10 years of energy prices fixed. The purpose was to provide non-utility developers with the long-term price certainty necessary to obtain financing for capital-intensive projects. Energy payments during the first 10 years were based on projections of future oil prices, which were expected to significantly increase (Meade and Porter, 1987).

The combination of federal and California incentives and innovative state regulation launched the wind energy industry in the US, and gave California the short-lived title of having the most installed wind capacity in the world. Wind development increased from 10MW in 1981 to a cumulative total of 1039MW in 1985, all located in California (AWEA, 2003). By the early 1990s, as most of the remaining ISO4 contracts were built out, California had about 1700MW of wind projects in place.

The demise of ISO4: Bust follows boom

ISO4 contracts did not last long. Oil prices did not increase as forecast but instead tumbled in 1985. That year, the California Public Utilities Commission (CPUC) suspended the ISO4 contract, and gave those that had signed these contracts five years to bring the projects online, or the contracts would be nullified. Although some criticize the ISO4 contracts for being overly generous to non-utility generators, the ISO4 payments are actually lower than utility-owned resources that were brought online during the same period, according to a study from the CPUC (CEC/CPUC, 1988). The ISO4 contracts effectively served as a commercialization venture for non-fossil resource development in California and played a major role in that state's worldwide leadership in the development of wind and other renewable resources.

That said, the discontinuance of the ISO4 contracts and the expiration of the federal tax incentives for wind soon after were major blows to the US wind industry. As a result wind development in the US stagnated and most of the nascent wind turbine manufacturers in the country went out of business as less than 500MW of wind was developed between 1987 and 1998.

What happened in other states?

Few states followed California's lead in opening markets for renewables. Even though federal tax credits were available across the country until 1986, and quite a few other states offered tax credits as well, California's regulatory treatment was uniquely effective at nurturing renewables development. No other state saw such a strong and broad-based growth of all renewable technologies.

The state of New York made an effort to encourage independent power with its famous 'six-cent' law, which paid independent generators a standard rate of six

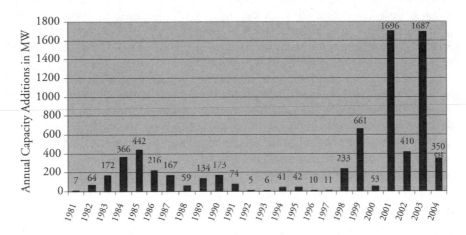

Source: American Wind Energy Association

Figure 7.1 *Year-by-year installed wind capacity in the US*

US cents per kWh over the course of the contract. Unfortunately, given their emerging status and still high cost, the rate was not sufficient to support renewables at that time. Instead it resulted in a rush of co-generation and overwhelmed the state's utilities with what turned out to be relatively high-cost power as the wholesale price of electricity trended down over the course of the next decade. The state of Maine also enacted high avoided-cost prices as a means of encouraging renewables. A number of biomass and small hydro projects were developed in Maine, but similar to New York, the power from these projects turned out to be relatively high cost as wholesale prices declined.

The avoided cost provisions of PURPA soon came to serve as an effective barrier to new renewables deployment. The reality was that renewables, in their still emerging state, simply were not in a position to compete head to head with coal or gas-fired power in most conventional avoided-cost calculations. Thus PURPA soon ceased to be an effective spur to renewables development. The only way new renewables projects were going to move forward was through a strong commitment from state policy-makers to diversify the utility portfolio with cleaner technologies. Very few states, with the exception of Iowa in 1983 and Minnesota in 1994, were willing to push such requirements.[2]

The Wind Production Tax Credit

The Production Tax Credit (PTC) for wind and closed-loop biomass was enacted with the Energy Policy Act of 1992 and went into effect on 1 January 1994. The PTC provides an inflation-adjusted tax credit; in 2005 this was 1.8 US cents for every kWh of production from eligible facilities for 10 years.

The PTC has expired twice for brief periods of time, to be renewed only for two-year periods, thus setting up the boom–bust cycle that has made sustained development of the wind industry difficult. The PTC expired again at the end of 2003, once again providing an example of this boom–bust cycle. In 2003, 1687MW of wind was installed in the US, but only 389MW was installed in 2004 given the unavailability of the PTC (AWEA, 2004). In 2005, Congress extended the PTC for two more years, until the end of 2007.

The PTC allows a business with taxable income to claim a 1.8 cent tax credit for every kilowatt hour of generation from a wind turbine for the first 10 years of a project's operation. The tax credit, along with power sale revenues, makes for an attractive revenue stream and helps wind projects to be financed. Because many wind companies do not have sufficient taxable income to take full advantage of the PTC, they often sell their wind projects to companies that have more taxable income and can take full advantage of the PTC.

Mixed blessings of the PTC

Many entities interested in wind development have difficulty taking advantage of the PTC. Most wind developers are not profitable enough to utilize the credits and even many large, profitable companies do not have the required long-term clarity regarding their tax liability. Municipal utilities and rural electric cooperatives, being not-for-profit entities, are also unable to utilize the PTC. Thus, the market for wind has become increasingly dominated by a few large companies such as FPL Energy and Shell that have a large 'tax appetite'. This is quite a different experience from other countries where the renewable industry has been built up by a new breed of players in the power sector.

The PTC has been both a blessing and a curse for the US wind industry. It is a significant incentive, so the timing of its expiration and extension has come to dominate the planning and financing of wind projects. Loss of the PTC would substantially reduce the size of the wind market in the US, but it has also added substantial costs to development because of some of its inefficiencies and the on-again, off-again pattern of expiration and extension.

The PTC is not a market maker

While the PTC is significant, it is not a market maker in the sense of the Electricity Feed Law or other policies such as a Renewable Portfolio Standard, which are both discussed later in the paper.

The proof is always in the outcome, not the policy. This can be seen in the fact that only 231MW of wind was developed in the first five years of the credit availability, the years 1994–98. We can also see the strong dependence of the proportion of development in the US that has been driven by state mandates. For example, just over 1000MW of the 1687MW of new wind capacity brought online in the US in 2003 was installed in a state with a Renewable Portfolio Standard (RPS) or was due in part to an RPS in a nearby state.

Given these limitations, it is clear that other policies deserve consideration to either supplement or supplant the PTC.

The RPS: a Renewable Energy Policy for Competitive Electricity Markets

The context of competitive power markets

State policy-makers became less willing to actively promote renewables in the early to mid-1990s as it became clear that the electric industry was moving towards a more competitive model. Using PURPA as a vehicle to advance renewables became even less viable as utilities and their regulators became increasingly nervous about the higher upfront costs of renewables.

At that point, the American Wind Energy Association (AWEA) began looking for a new policy approach that was more compatible with the emerging competitive market. Starting in 1993, AWEA began serious efforts to define a new policy mechanism. After months of dialogue with other renewable energy advocates, AWEA, in collaboration with the Union of Concerned Scientists (UCS), began developing the basic outlines of the new policy – the Renewable Portfolio Standard.

The RPS components

An April 1995 AWEA fact sheet described most of the major components of the new policy, which are as follows.

- Competitively neutral: Applies to all competitors equally.
- Market-driven: 'Instead of tax/subsidy schemes and bureaucratic implementation, the RPS would rely on the free market to ensure that renewables are developed in the most economical way' (AWEA, 1995).
- Flexibility: Modelled on the Clean Air Act Amendments of 1990, the RPS introduced the concept of tradable Renewable Energy Credits (RECs) as a means of encouraging the most cost-effective compliance with the standard.
- Simplicity: Simple to implement with minimal government involvement which is limited to monitoring compliance and facilitating trading.
- Complements research, development and deployment (RD&D) goals: Could use state or federal RD&D funds to leverage emerging renewable technologies such as PV, small wind or solar thermal into the market.
- Stable and consistent renewables policy: The RPS provides a long-term, predictable market for renewables 'which can be expected to drive technology costs down rapidly' (AWEA, 1995).

The critical role of the states

The RPS approach has been applied nationally in Australia, the UK and now Japan. However, in the US, the RPS has never been adopted on a national basis. In 1996 it was initially included in federal restructuring legislation by Representative Dan Schaefer (Republican, state of Colorado), who chaired the House of Representatives Energy and Power Subcommittee at that time. It was later incorporated into Clinton administration restructuring proposals and was passed by the Senate in 2002 and in 2004. However, Congress has never succeeded in including the concept in federal law.

The great success of the RPS concept in the US has been at the state level, with 21 states (and the District of Columbia) having established such a policy. Texas is probably the best example of the RPS and has been a testing ground for an international policy audience.

The Texas Renewable Portfolio Standard[3]

Implementation

Although the RPS is simple in concept, implementing it so as to achieve the desired outcome of a growing market for renewable energy requires attention to detail. There are a number of RPS implementation details that are important to get right; otherwise the desired economic and environmental benefits from an RPS will not be realized. Since the most successful RPS in the US by far has been the Texas RPS, it is worth examining why it is so successful.

Table 7.1 *The Texas RPS requirements*

Year[4]	Required MW of New Renewables
2003	400
2005	850
2007	1400
2009	2000

Source: American Wind Energy Association

The Texas Legislature approved the RPS in 1999 as part of electric utility restructuring legislation. The Texas RPS requires 2880MW of renewable energy by 1 January 2009, with 2000MW of that to come from new renewable energy projects (see Table 7.1). Subsequently, the Texas Public Utilities Commission (PUC) converted the capacity goals into energy requirements, using an initial capacity factor of 35 per cent and adjusting it over time based on actual plant performance. The PUC did its part by speedily implementing the RPS provisions, with implementing regulations finalized just eight months after the RPS was passed.

What made the Texas RPS work?

The Texas RPS features many important elements that make it such an effective policy. These include:

- placing the RPS requirement on all electricity retailers serving competitive markets;
- an accessible REC trading system via the Internet;
- flexibility in meeting the RPS requirement, such as allowing the banking of RECs for up to two years and allowing electricity retailers to borrow up to 5 per cent of their RPS obligations;
- a strong compliance regime that includes a penalty for non-compliance that is the lesser of 5 US cents/kWh or 200 per cent of the mean REC trade value[5]; and
- a careful definition of the renewable technologies. By contrast, other states such as Maine had such a broad definition, including a number of non-renewable technologies, that the state RPS had virtually no impact on the market for renewables.

The impact of the RPS

All told, the Texas RPS represents about 3 per cent of electricity consumption in Texas. Although that number appears modest, the Texas RPS has significantly increased renewable energy capacity in Texas, and is one of the most aggressive state renewable energy policies in the US in terms of capacity additions. The large RPS in Texas allows wind developers to build larger projects and gain economies of scale. With a high-quality wind resource that is second only in the US to North

Table 7.2 *RPS Obligations and wind contracts for retail suppliers*

Electricity supplier	Approx. 2003 RPS Obligation (MW)	2001 Wind contracts (MW)
TXU	170	353
Reliant	140	208
AEP	0	0
Entergy	0	0
Excel-SPS	40	80
TNP	2	3
Enron	15	130
Other new players	33	?
Austin	0	80
LCRA	0	50
San Antonio	0	25
El Paso	0	1
TOTAL	400	930

Source: Wiser and Langniss (2001)

Dakota, along with the availability of the federal PTC, wind power projects in Texas can deliver power for less than 3 US cents/kWh. Table 7.2 illustrates the wind capacity that has come online in Texas.

Market activity has already made Texas a leading state in the US in terms of installed wind capacity. The Electric Reliability Council of Texas (ERCOT) states that there are 1186MW of renewable energy capacity within ERCOT, and of that 1139MW is from wind (ERCOT, 2004). Thus, the Texas RPS requirement was met through 2005 with the renewable resources that had already been brought online (Wiser and Langniss, 2001). The Texas PUC believes the full 2000MW requirement could be met well before the 2009 deadline (Texas PUC, 2003a). In 2005, Texas increased their RPS to 5880MW by 2015, with a target of 10,880MW by 2025.

RPS and contract periods

Helping keep RPS compliance costs low in Texas is the willingness of electric retailers to enter into long-term contracts of 10–25 years for the electricity and RECs. Long-term contracts provide renewable energy developers with a steady revenue stream and make financing possible. This, in turn, allows electricity retailers to receive renewable electric power at stable prices.

RPS and transmission

Despite these successes, the Texas RPS has recently run into some obstacles. Most of the state's wind capacity is in west Texas, where there is a low population and consequently an undersized transmission system. The speed with which wind systems were installed – about one year – has overwhelmed transmission planners, who typically need five to eight years to install new transmission. As a result, wind generation has been curtailed at times in Texas and concern over these curtailments briefly drove up REC prices in the summer of 2002.

The market and regulatory response has been swift, suggesting an eventual resolution of these problems. ERCOT has recently approved several major new transmission improvements and additions in west Texas that should eventually help alleviate most transmission issues. In addition, electricity retailers are looking at other sites in central Texas that, while perhaps offering not as high a quality of wind resource, are not likely to face transmission constraints.

RPS and certainty

Even in some states with well-developed RPS policies, the unavailability of long-term contracts has made it difficult for new renewable energy projects to be built. For this reason, at least in part, more recent RPS policies in California and Nevada require that long-term contracts of at least 10 years be signed with renewable energy generators.

Meanwhile ... Wind Energy in Europe

The world's fastest growing wind energy market is in Europe. Total installed wind capacity in Europe at the end of 2004 was just over 23GW, with 85 per cent of the capacity in Denmark, Germany and Spain (see Table 7.3). The rapid growth prompted the European Wind Energy Association to estimate that 100GW of wind could be installed in Europe by 2010 (EWEA, 2002).

The Electricity Feed Law: A fixed-price approach

A major driver of wind market growth in Europe has been the Electricity Feed Law (EFL), adopted in Denmark, Germany, Italy and Spain. In contrast to the RPS, which takes the approach of a fixed quantity of generation by a certain date, the EFL relies upon a fixed-price approach to renewable energy development.

The German parliament unanimously approved the EFL (known as the Stromeinspeisungsgesetz in Germany) to invest in climate protection, preserve conventional energy resources and promote renewable energy resources. The German EFL took effect in 1991 and requires utilities to buy renewable energy from independent power producers at a government-defined price. The Renewable Energy Feed-In Tariffs are linked to the average electricity retail rates for all electricity consumers in Germany. Wind and solar received 90 per cent of the average retail electricity rate; biomass and hydro under 500kW received 80 per cent; and small hydro up to 5MW received 65 per cent (Wagner, 1999).

However, difficulties arose in marrying the EFL with the EU-wide implementation of competition in the power markets, with several companies taking legal action on the grounds that they were being put in a position of competitive

Table 7.3 *European wind energy capacity by country*

European wind energy markets (by installed capacity in MW)	2004 Additions	2004 Year end total
Germany	2037	16,629
Spain	2065	8263
Denmark	9	3117
Italy	221	1125
Netherlands	197	1078
UK	240	888
Austria	192	606
Portugal	226	522
Greece	90	465
Sweden	43	442
France	138	386
Ireland	148	339
Belgium	28	95
Finland	30	82
Luxembourg	14	35
EU Total	5703	34,205

Source: AWEA and EWEA (2005)

disadvantage. There were also complaints that while the price for wind energy was fixed, the cost of the technology was becoming cheaper and might lead to excessive profits at the taxpayers' expense.

Refining the EFL: The Renewable Energy Law

Germany is in the midst of changing its Renewable Energy Law (REL), which was specifically applied to wind, solar and geothermal facilities; for hydro, landfill gas and sewage or mine gas facilities under 5MW; and biomass facilities under 20MW (Knight, 2004a).

As with the old law, the REL requires that renewable energy be purchased at specified tariff rates, but this time with the amount of renewable energy distributed equally among all electricity suppliers. This avoided the competitive disadvantage of some utilities located in areas of high renewable energy resource.

As proposed, the rate Germany will pay for a new wind project is half of what it was in the early 1990s and will decline by an additional 36 per cent over the next decade (Knight, 2004b). The revised law continues to classify wind as priority power and requires grid operators to upgrade the grid when necessary for interconnecting a new wind project, as long as the cost is considered to be economically reasonable. The grid operator must provide documentation within four weeks if they refuse to interconnect a wind project because of lack of available grid capacity. Every two years the German parliament re-evaluates the REL, and potentially the rates of tariff, based on a report submitted by the German Ministries of Economics and Technology in consultation with the German Ministries of Environment and Agriculture.

The Danish example

Overall, Danish policies have been effective, resulting in the largest wind turbine manufacturing industry in the world. The growth of wind energy in Denmark was fuelled by policies similar to the REL. These allowed wind power generators to receive payments of 85 per cent of retail electricity rates, as well as companion policies such as capital subsidies, tax incentives, low-cost financing and R&D funding (Wiser et al, 2002). Denmark has changed this policy extensively, but its fixed price period was instrumental in getting the Danish wind industry off the ground.

Denmark received 27 per cent of its electricity consumption from renewable energy by 2003, exceeding the goal of 20 per cent set for that year by the 1996 Danish Energy 21 plan (Danish Wind Energy Association, 2002). Another 400MW of offshore wind and 350MW of onshore wind are expected over the next five years (Moller, 2004).

Implications and Summary

In recent years, EFL policies have come under political pressure, as the European Union has been more supportive of a European-wide renewables requirement instead of relying on EFL policies. This has set off a debate in Europe over whether fixed-price policies (such as the EFL) are preferable to fixed-quantity systems (such as the RPS). Both the EFL and the RPS are effective market stimulants for renewable energy if designed properly. Both policies have had successes and failures, largely attributable to how each policy was designed in specific circumstances.

The EFL shares common characteristics with California's implementation of PURPA in the 1980s, with an above-market fixed price paid for each kWh for a period of time. Both the ISO4 and the EFL, in conjunction with other incentives, provided the impetus for launching the US and European wind industries, respectively. The primary difference is that the price paid under the EFL is based on a percentage of the retail price, while the price paid under the ISO4 was based on state and utility forecasts of future oil prices.

The EFL can provide great stimulus to renewable energy development, and it is also relatively easy to administer. A critical factor for the EFL is setting the price. If the price is too low, the market incentive is insufficient and little renewable energy development will take place. Conversely, if the price is set too high, political and public support may crumble if costs are perceived as excessive. Investors may demand a risk premium if they perceive any uncertainty over whether EFL policies will continue and for how long. Renewable energy development may also stop during political negotiations to change a policy, as it did in Denmark during electricity reform negotiations (Krohn, 2000).

The EFL may also affect some market participants more than others. For instance, utilities in renewable-rich areas may incur most of the cost impacts of EFL policies. This has been an issue in Germany, where some utilities in northern Germany had greater cost impacts from EFL policies because of their proximity to good wind resources. Germany is now spreading some of these costs to other participants, but this offsets one of the main advantages of EFL policies, namely its ease of administration.

For at least some of these reasons, a fixed price type of policy such as the EFL is politically untenable in the US, where great emphasis is placed on market competition to serve public demand. Given large regional differences in the wholesale price of electricity and the cost-competitiveness of renewables, a single national price for renewable energy is impractical, and setting regional or utility-specific fixed prices is more workable but could become process heavy and bureaucratic. The passage of time has not dimmed utility memories of the negative experience of New York's six-cent law. Fixed prices simply have difficulty keeping up with rapid evolution of technology and steadily improving economics, whether of wind or competing technologies such as gas-fired generation. Controversy still rages over the effects of PURPA, even though state implementation of PURPA occurred nearly 20 years ago. Competition-based policies such as the RPS have greater public and political support. Even with its implementation in a relatively few states, the RPS has had the salutary effect of helping to drive the cost of wind-generated electricity consistently lower. In the end, wind's market share will be determined by how well it competes economically with other electric resource options and the extent to which RPS policies have helped facilitate lower-cost renewable resources.

A well-designed RPS can provide significant market stimulus to the wind energy industry for a prolonged period, allowing for sufficient and sustainable market growth over time. While the RPS is simple in concept, numerous implementation details must be done correctly to get the desired impact of the RPS. These details include:

- applying the RPS to at least the majority, if not all, of electricity retailers;
- providing a steady but not insurmountable increase in the level of renewables over time;
- clear start and end dates;
- clear resource and geographic eligibility definitions and rules;
- flexibility mechanisms to ease compliance;
- a renewable energy credit trading system; and
- an emphasis on long-term contracting.[6]

That leads us to our central conclusion. Clearly, different policy approaches can work in helping to facilitate a market for an emerging renewable technology such as wind, and both the European EFL and the RPS have demonstrated their effectiveness. The key to a booming European wind market, however, has been policies that have been relatively stable and long term, providing significant incentives for wind that have now been in place for at least a decade. In contrast, federal renewable energy policy in the US has subjected the wind industry to a policy roller-coaster ride over the past two decades, as the year-to-year chart of installed capacity demonstrates (Figure 7.1). Such inconsistent policy support provides a hostile climate in which to nurture a manufacturing industry.

How do the differences in US and European policy play out in terms of the impact they have had on the wind industry? Perhaps nowhere can this be better exemplified than by looking at the location of the world's largest wind turbine manufacturers. Seven of the ten largest wind turbine manufacturers are in Denmark, Germany or Spain, the three countries with the strongest and most consistent market incentives. Despite the status of the US wind market as second largest in the world, only one of the ten manufacturers is from the US. If the US hopes to foster a larger role in this multi-billion dollar industry, it will have to provide more effective and consistent policy support than it has provided in the past.

Notes

1   Also in 1980, Congress enacted tax incentives of 10 per cent for business investments in biomass and 11 per cent for hydro projects of 25MW or less. These incentives have since expired.
2   Iowa passed a law requiring its electric utilities to install 105 average megawatts of new renewable energy generation. After considerable legal wrangling and delays, 250MW of wind was installed in the late 1990s. Minnesota required 400MW of wind and 125MW of biomass (reduced by the Minnesota legislature to 110MW) by 2002, and an additional 400MW of wind by 2012 (later changed to 2006 as part of a regulatory settlement).
3   Unless otherwise stated, the source for this section is Wiser and Langniss (2001).
4   'Year' refers to 1 January of the given year.
5   Like many states that have restructured their electric power industry, municipal utilities and rural electric co-operatives in Texas do not have to open their markets to retail competition but may do so voluntarily. If they voluntarily open their markets, they must comply with the RPS. These entities account for about 20 per cent of retail electric sales in Texas. Also, the Texas PUC recently raised the REC deficit allowance from 5 per cent to 10 per cent, and the compliance grace period from three months to six months (Texas PUC, 2003b).
6   Even in some states with well-developed RPS policies, the unavailability of long-term contracts has made it difficult for new renewable energy projects to be built.

References

American Wind Energy Association (AWEA) (1995) *A Renewables Portfolio Standard*, AWEA, Washington, DC

AWEA (2003) *Wind Power: US Installed Capacity (Megawatts) 1981–2002*, www.awea.org/faq/inst cap.html, accessed March 2005

AWEA (2004) *First Quarter Report: Wind Industry Trade Group Sees Little to No Growth in 2004, Following Near Record Expansion in 2003*, 12 May, www.awea.org/news/news040512.1qt.html, accessed March 2005

AWEA and European Wind Energy Association (EWEA) (2004) *Global Wind Power Growth Continues to Strengthen*, 20 March, www.awea.org/news/news040310glo.html, accessed March 2005

California Energy Commission (CEC) and California Public Utilities Commission (CPUC) (1988) *Final Report to the Legislature on Joint CEC/CPUC Hearings on Excess Electrical Generating Capacity*, P150-87-002

Danish Wind Energy Association (2002) *Wind Energy News Archives 2002*, www.windpower.org/news/archive6.htm

Electric Reliability Council of Texas (ERCOT) (2004) *Existing/New REC Capacity Report*

EWEA (2002) *European Wind Energy Achieves 40 per cent Growth Rate*, 3 November, www.ewea.org/doc/13-11-02%20European%20wind%20energy%20achieves%2040%25%20growth%20rate.pdf, accessed March 2005

Knight, S. (2004a) 'Revised German Wind Law Delayed', *Windpower Monthly*, June, p32

Knight, S. (2004b) 'Downward Pressure on German Prices' *Windpower Monthly*, May, p34, p38

Krohn, S. (2000) *Renewables in the EU Single Market – An Economic and a Policy Analysis*, Paper presented at the European Small Hydro Conference, May, www.windpower.org/articles/reneweu.htm, accessed May 2003

Moller, T. (2004) 'Denmark Back on Track' *Windpower Monthly*, May, p6

Meade, W. and Porter, K. (1987) *Trends in State Utility Regulation Affecting Renewable Energy*, Washington, Renewable Energy Institute

Texas PUC (2003a) *Report to the 78th Texas Legislature: Scope of Competition in Electric Markets in Texas*, January, www.puc.state.tx.us/electric/reports/scope/2003/2003scope_elec.pdf, accessed March 2005

Texas PUC (2003b) *Order Adopting Amendments to §25.173 As Approved at the January 30, 2003, Open Meeting*, 6 February, www.puc.state.tx.us/rules/rulemake/26848/26848adt.pdf, accessed March 2005

Wagner, A. (1999) *Wind Energy Development in Germany: A Joint Success Story of the Industry, Owners and Politics*, www.boell.de/de/04_thema/840.html, accessed March 2005

Wiser, R. and Langniss, O. (2001) *The Renewables Portfolio Standard in Texas: An Early Assessment*, Ernest Orlando Lawrence Berkeley National Laboratory, LBNL-49106, November, eetd.lbl.gov/ea/EMS/reports/49107.pdf, accessed March 2005

Wiser, R., Hamrin, J. and Wingate, M. (2002) *Renewable Energy Policy Options for China: A Comparison of Renewable Portfolio Standards, Feed-In Tariffs, and Tendering Policies*, San Francisco, Center for Resource Solutions, June

8
Development of Renewable Energy in India: An Industry Perspective

Rakesh Bakshi

Introduction: A Country of Rapid Change

Power is a critical infrastructure input for the development and growth of the economy of a country. Since independence in 1947, the installed power capacity in India has increased from 1.4 gigawatts (GW) to over 100GW and more than 500,000 villages have been electrified to date in India. Nevertheless, India's per capita energy consumption is relatively low compared to developed countries and a large number of villages still have no access to electricity. With a growing population of over one billion and an ongoing path of industrialization, India is already and will continue to be one of the world's largest markets for new energy in the world.

The predicament of the Indian power consumer

Before looking at new energy sources in India, we must spend a little time looking at the lot of the Indian energy consumer today. The electricity supply is an excellent example.

Today, the end users of electricity, such as households, agricultural farms, commercial establishments and industries, are confronted with frequent power cuts, both scheduled and unscheduled. Erratic voltage levels and wide fluctuations in the frequency of supply have added to the power problems of consumers. The major causes behind the inadequate, erratic and unreliable power supply in India can be summarized as:

- inadequate power generation capacity;
- a lack of optimum utilization of the existing generation capacity;
- inadequate inter-regional transmission links;
- inadequate and ageing transmission and distribution networks leading to power cuts and local failures/faults;

- unauthorized tapping of power;
- skewed tariff structures;
- the slow pace of rural electrification; and
- Inefficient use of electricity by the end consumer.

As a consequence, today's consumers are resorting to stand-alone and back-up solutions whether inside or outside the power grids. These captive power supply arrangements include diesel generator sets or battery banks, inverters and power stabilizers of various capacities ranging from 250 watts (household) to hundreds of megawatts (industry). So even though one might consider India an industrializing country with low-level affluence, the money spent by the domestic and industrial sectors on these standby power supply arrangements may be among the highest in the world. This leaves some room for opportunity, especially for renewables where price is always a critical issue.

Strengths and opportunities in the Indian power sector

Again before looking at renewable energy in particular it is worth considering the energy sources to which India has access. It is also worth looking at some of the elements that are creating and attracting global energy players (both renewable and conventional) to set up business in India:

- abundant coal reserves (enough to last at least 200 years);
- vast (large-scale) hydroelectric potential (150,000 MW);
- large new renewable energy potential including 22,000MW equivalent of biomass waste from agriculture, very high solar insolation levels for electricity or heat production, 45,000MW of possible wind capacity, 15,000MW of small hydro and 2500MW equivalent of waste;
- large and increasingly skilled pool of highly skilled technical personnel;
- emergence of strong and globally comparable central utilities (NTPC, Power Grid);
- enabling framework for private investors;
- well laid out mechanism for dispute resolution;
- political consensus on reforms; and
- one of the largest future power markets in the world.

Meeting the challenge: The objectives for the Indian power sector

In order to address the current limitations and to take advantage of the assets discussed above, the following three objectives are seen by most commentators as the key medium-term goals for the Indian power sector as it expands to meet the needs of the population:

1   to provide power on demand by 2012;
2   to provide reliable and quality power at an economic price; and
3   to achieve environmentally sustainable power development.

For developed country markets the second two points, price and environment, are usually the focus of discussion, with the addition of energy security. In India the difference is that the provision of commercial energy is not yet secure. As discussed above, electricity is still intermittent and not yet of high quality in many areas. So the first point might be thought of as equivalent to achieving secure commercial energy in India.

Achievements so far in harnessing India's renewable resources

The contribution of renewables to total electricity generation had reached over 3500MW up to 31 March 2002, representing approximately 3.3 per cent of the total installed capacity in India. Considering the estimated total renewable potential of 100,000MW, the achievements so far leave much unharnessed potential. The Ministry of Non-conventional Energy Sources (MNES) has been preparing a comprehensive perspective plan under which 10 per cent of the total additional power generated by the year 2012 would be from the renewable sector. It is estimated that in the next 10 years India would require a capacity addition of around 100,000MW and as such the contribution to the Indian grids from the NRSE sector should be substantial.

Three Key Drivers for Renewable Energy in India

India, like China, is interesting because it is both a large industrializing country which already has considerable infrastructure and demands, but it is also a socially

Table 8.1 *The potential and achievement of various New and Renewable Sources of Energy (NRSE) technologies in India*

Sources/systems	Potential	Achieved
Biomass power	19,500MW	381 MW
Biomass gasifiers		51 MW
Biogas plants (no.)	12 million	3.26 million
Improved chulhas (no.)	120 million	34.3 million
Solar photovoltaics		85 MW
		(includes 30MW of exports)
Solar thermal systems (collector area m²)	140 million	0.60 million
Solar power		2.0MW
Wind power	45,000MW	1617MW
Small hydropower	15,000MW	1438MW
Waste to energy	1700 MW	22 MW

Note: Achieved data are as at 31 March 2002.

developing country with services for home or business which are made possible with commercial energy yet to be supplied to large numbers of people.

The key reasons that have necessitated the wider use of renewable energy technologies in India, are as follows:

- rapidly growing energy demands and inability of the conventional systems of power generation to meet this challenge in an equitable and sustainable manner;
- local and global environmental degradation as a consequence of generating power with conventional fossil fuels;
- the imperative need for meeting energy needs of the unserved and under-served populations in far-flung rural and remote areas;
- depletion of indigenous fossil fuel reserves.

The need for new power generation

Based on the projections of demand made in the 16th Electric Power Supply survey, additional generation capacity of over 100,000MW needs to be added to ensure power on demand by 2012. This amounts to nearly doubling the existing capacity of about 100,000MW and it is estimated that for building the additional power capacity, associated transmission and distribution infrastructure, nearly 8000 billion rupees (US$175 billion) would have to be invested.

This is of course important for renewables, as meeting the additional require-ment for energy and power entails taking advantage of all economically viable sources of energy and working out a suitable strategy of energy mix to gain the optimum advantage. Indigenous renewable sources clearly have a role.

Energy security

For any growing economy, energy security is crucial to assure its smooth devel-opment. Moreover the economy will also be very sensitive to the cost of energy imports. The mix of India's current installed capacity of over 100GW of power is approximately 70 per cent thermal, 24 per cent hydro, 3 per cent non-hydro renewables and 2 per cent nuclear.

India is a net importer of energy, particularly oil, accounting for around 30 per cent of total energy consumption, which in turn constitutes nearly 20 per cent of the total imports of the country. Oil imports are of course likely to increase with growth.

Fossil fuels are expected to fuel the economic development process of the majority of countries for another 10–20 years. However, it is anticipated by sev-eral energy scenarios that fossil fuels may reach their maximum potential during the period 2020–2050, after which the cost of generating electricity with fossil fuels would exceed that of other sources of energy generation, notwithstanding environmental constraints.

So for a country like India which is installing the bulk of its energy supply this century, both the short-term energy security exposure to importing energy and the long-term pricing expectations merit early consideration of alternative indigenous sources of energy. Renewables provide several good candidates that could address energy security issues.

Environmental protection

The population of India – the world's second most populous nation – grew from 300 million in 1947 to more than one billion today. This rapidly increasing population, along with progress in economic development, has posed other problems such as environmental degradation in addition to rapidly depleting natural resources. Domestic fossil fuels are not only finite, but their use leads to large-scale atmospheric carbon emissions causing global warming.

Domestic concerns about air quality are one side of the equation, but with such a populous nation, the trajectory of the Indian energy sector in terms of greenhouse gas (GHG) emissions will be of global significance. Hence, there is an urgent need to use more environmentally sound and less hazardous technologies to counter the menace of environmental pollution. The increase in population coupled with rapid industralization will result in a significant rise in per capita emissions during the next 10–20 years. Japan has a per capita emission intensity that is over 10 times that of India; it is double again for the US. So, if the future energy of Indians is met in the same ways as it is in Japan or the US, there are serious consequences. The solution is a shift towards environmentally benign energy resources that can fuel the developmental process without adding to the harmful emissions not only at the global level but also at the local and regional levels.

Energy in remote areas

About 70 per cent of India's one billion citizens live in rural areas and yet approximately 18,000 villages have no electricity. This not only affects the quality of life available to those people, but it also limits the type of commercial activity that is available to them and therefore their levels of financial security.

Equally important and limiting is the need for cooking fuels. With an expanding population and limited supplies of local fuels, the result is that a family member may have to walk many kilometres each day to collect the required fuels to cook. Again this not only impacts wellbeing, but also severely limits the ability of family members to be economically active and therefore reduce the levels of financial poverty. Renewable energy sources are often well suited to remote areas or stand-alone systems, and would therefore offer suitable solutions to address some of these economic problems.

The Evolution of Renewable Energy Policy-Making in India

The renewable policy context

Two major issues have been responsible for putting renewable energy sources in the right perspective in India; the widening gap between energy consumption and supply and the resultant polluting emissions generated by using conventional fossil fuels. Furthermore, renewable energy technologies have an important role to play in the socio-economic development of the communities in developing countries, including India.

India was among the few first countries in the world to have launched a major programme for harnessing renewable energy. By the 1990s the government of India came to recognize the importance of renewable energy as a means of decentralized energy systems which could meet the requirements of far-flung rural areas. The government of India had given a new thrust to its efforts at the beginning of the Eighth Five-Year plan (1992–97) which included the use of renewable energy technologies for power generation.

Over the past two decades, several renewable energy technologies have attained technological maturity in India, leading to rapid commercialization. By giving greater importance to developing market linkages, as well as involving the private sector in these ventures by providing fiscal and tax incentives, the government of India has sought to accelerate the pace of development.

Thus the importance of an increasing use of renewable energy sources in the transition to a sustainable energy base continues to be appreciated. However, it must be emphasized that there is an entirely different set of issues that India grapples with, and with which renewable energy systems can assist. Renewable energy not only augments energy generation and GHG emissions mitigation but also contributes to improvements in the local environment, afforestation, drought control, energy conservation, employment generation, upgrading of health and hygiene, social welfare, security of drinking water, increased agricultural yield and production of bio-fertilizers. Hence, policy-makers recognize that in order to achieve sustainable development as a whole, the growing energy requirements in India can be met by a judicious deployment of renewable energy resources if a suitable pattern of energy mix is evolved.

The three distinctive stages of renewable energy development

When looking at the historical development of renewable energy in India we can see three distinct stages:

1 In the first stage, from the late 1970s to the early 1980s, the thrust of the national effort in renewables was directed towards capacity building and R&D, largely in national laboratories and educational institutions.
2 The second stage, from the early 1980s to the end of that decade, witnessed a major expansion of activity with the emphasis on large-scale demonstration and subsidy-driven extension activities, mainly in the field of biogas, improved cooking stoves and solar energy.
3 In the third and current stage, beginning in the early 1990s, the emphasis has been more on the application of matured technologies for power generation, based on wind, small hydro, biogas, co-generation and other biomass systems, as well as for industrial applications of solar and other forms of energy. There is also a gradual shift from the subsidy-driven mode to commercially driven activity.

The lineage of institutional frameworks

The relevance of the increasing need to use renewable energy sources as a sustainable energy base was recognized in India even as early as the 1970s. One of the significant successes in India was the establishment of government agencies empowered to make renewable energy production happen. In 1981, the government of India set up a commission under the name of the Commission for Additional Sources of Energy (CASE) to promote the development of renewable energy technologies for use in different sectors.

In 1982, this process was taken a step further when the government of India created a separate department – the Department of Non-conventional Energy Sources (DNES) in the Ministry of Energy and charged it with promoting the development of non-conventional energy sources.

A decade later in 1992, the DNES was upgraded to the status of a fully fledged ministry with the title Ministry of Non-conventional Energy Sources (MNES), with a mandate to develop all areas of renewable energy. This has earned India the distinction of possibly being the only country in the world to have an exclusive ministry dealing with new and renewable energy sources.

The power of these institutional steps results in policy setting from the very top of government and their impact cannot be overstated. They are possibly the main reason that a country that is still very much in the process of development and industrialization has been able to leapfrog to the international fore in a set of very new and advanced technologies.

Support, Targets and Instruments

National and regional resource assessments

As we see in other areas of renewable energy industry development, the cost of large-scale resource assessment is prohibitively high for individual companies and is something that is more appropriately undertaken for the industry as a whole. For this reason the Indian government has taken a strategic approach to resources assessment for specific renewables. For example, the MNES has undertaken a Biomass Resources Assessment Programme covering all the states and union territories with the objective of providing inputs for a Biomass Resource Atlas for India.

Small hydro is another excellent example. India's first small hydro project of 130KW was commissioned in the hills of Darjeeling in 1897. Small hydro is environmentally benign, operationally flexible and suitable for peak period support to the local grid as well as for stand-alone applications in isolated areas. India has one of the world's largest irrigation canal networks with thousands of dams and the total potential of small hydropower projects of sizes up to 25MW has been assessed at 15,000MW. As of 31 March 2002, about 425 projects with capacity up to 25MW (with an aggregate capacity of 1438MW) had been commissioned. Another 187 projects with capacity up to 25MW (with an aggregate capacity of approximately 521MW) were under implementation. Various state governments have shown their interest in establishing small hydropower projects in their respective states. A key intervention by the MNES has been to create a database of the sites potentially suitable for small, mini and micro hydro projects by collecting information from various sources including the state governments. The database created by MNES now includes 4096 sites with a potential capacity aggregating 10,071MW.

Target setting and policy frameworks

Most successful industry development processes start with targets, and Indian renewable energy development is no exception. The government of India initially announced its draft renewable energy policy, wherein it is envisaged that 10 per cent of capacity additions will be met through renewable energy by 2012.

The role of renewable energy has been incorporated into the wider policy reforms currently underway. These include:

1 the unbundling and privatization of the electricity sector in India;
2 the new Electricity Bill, approved by the Union Cabinet of the government of India; and
3 formation of electricity regulators.

With an appropriate institutional framework and proper policies in place, the target renewable energy market will be worth 600 billion rupees (approximately US$13 billion) by 2012.

Targeted markets and consumers

Given the role envisaged for some renewable energy technologies for development and improving the quality of life in rural areas, there have also been a number of carefully targeted programmes to assist in the deployment of the appropriate technologies relevant to family households.

For example, the MNES has been implementing two schemes: the National Project on Biogas Development (NPBD) and the National Programme on Improved Chulhas (solid fuel-burning stoves that use wood or dung) (NPIC) for family-type biogas plants and women's welfare biogas plants respectively. These are in addition to community- and institutional-based biogas plants.

Market development and industry development

The renewable energy market in India has passed through 100 billion rupees (approximately US$2.2 billion), and is growing at the rate of 15 per cent every year. The harnessing of renewable sources of energy in India constitutes a small but rapidly growing industry that is dominated by small and medium-sized enterprises. These companies are able to exploit not only the available human talent, but also adapt themselves rapidly to technological and market developments.

India's indigenous renewable energy technology development has generally followed a less than normal path-breaking route. Mostly, there has been steady technology growth in all the key sectors of renewable energy. Renewable industry products have followed a downward sloping cost curve, with a few selective breakthroughs oriented towards substantial cost reduction. However, most of the renewable energy systems and devices continue to be burdened with high initial costs, which many feel is the major deterrent to the growth of renewable energy.

Presently most manufacturing companies engaged in the manufacture of renewable energy products are relatively small enterprises and their capacity to mobilize resources to undertake technology upgrades is rather limited. Technology change also impacts on their capacity to develop an extensive network for product delivery as well as for proper after-sales maintenance. Addressing these issues and ensuring that medium and large industrial houses come forward and establish local production facilities to manufacture renewable energy products will be essential to the healthy growth of the industry.

Realizing the need for sustainable energy as the basis for sustainable economic development, the government of India has embarked upon programmes to deliver innovative financial and fiscal policies, financing instruments, institutional arrangements, human capacity development, information dissemination and industrial production.

The incentives currently extended to the renewable energy sector by the central government and/or various state governments in India include:

- providing budgetary resources from government for demonstration projects;
- loans on soft terms from the Indian Renewable Energy Development Agency and other financial institutions for private sector development of commercially viable renewable energy projects;
- leverage of external assistance from international bilateral agencies; and
- encouraging private investment through fiscal incentives, tax holidays, depreciation allowance, facilities for wheeling and banking of power, and ensuring remunerative returns for the power supplied to the grid.

The application of joint venture laws for technology transfer

Joint ventures have led to particular successes in establishing renewable energy industries in India. Led by huge international companies, such as BP, teaming up with huge indigenous companies, such as Tata, to manufacture solar photovoltaic systems, the model has provided a path for rapid technology transfer and the development of indigenous human and industrial capacity.

In India, joint ventures are possible not only in setting up the manufacturing of renewable energy products/devices but also for setting up renewable energy-based power generation projects. India offers a liberalized foreign investment policy, which allows foreign investment and transfer of technology through joint ventures. Investment proposals in the renewable energy sector with up to 74 per cent foreign equity participation in a joint venture qualify for automatic approval. Projects with 100 per cent foreign investment as equity in the renewable energy sector are permissible with the approval of the government of India, Foreign Investment Promotion Board (FIPB). Foreign investors can also set up a liaison office in India.

Industry- and technology-specific fiscal intervention

Subsequent to identifying appropriate national resources to be exploited or technologies to be applied, the government has chosen to enact very specific policies to target industry development.

The promotion of biomass-based power generation in India is being encouraged both at the state and central levels by a policy that includes fiscal incentives such as concessions on custom duties, exemptions from excise duty, tax holidays and accelerated depreciation for commercial projects.

For India's PV market, key drivers have been long-standing government programmes of subsidy, tax and financial incentives that began in the 1980s. Subsidies have accompanied the installation of most solar home systems, while loan and financing schemes have further supported private sector sales.

Incentives available for setting up wind power plants in India include:

- a 100 per cent depreciation in income tax assessment on investment on the capital equipment in the first year;
- a five-year tax holiday on income from the sale of generation;
- industry status, including capital subsidies in certain states;
- electricity banking and wheeling (moving) facilities;
- buy back of power generation by state electricity boards at remunerative price; and
- third-party direct sale of power generation in certain states.

These incentives coupled with the steep increase in demand for energy has led to an unprecedented enthusiasm by private companies to set up wind farms.

Accelerated technology development

During the past quarter century, a significant effort has gone into the development, trial and induction of a variety of renewable energy technologies for use in different sectors of the economy and sections of society in India. India, today, has one of the world's largest programmes for renewable energy technology development with activities covering all the major renewable energy sources of interest to India.

In many cases, there are technologies that India has been able to transfer from overseas and then build on domestically, such as wind turbine technology from Denmark and Germany. On the other hand, many renewable technologies under technical development are still in their infancy internationally, and may not have been as appropriate for the resources in other countries as they are in India. In either case it is important for government to help to prioritize public and private sector investments in the most viable resources and technologies.

An example of the success of such policies is the wind sector in India – one of the great success stories of the renewable energy development programmes in India. The Government of India, through the DNES and then the MNES took the decision at the beginning of the Seventh Five Year Plan (1987–1992) to import complete wind electric generators and install them as demonstration wind power projects with a view to creating awareness, providing operating experience and establishing the techno-economic feasibility of wind power generation. An additional aim was to study the various technical issues involved in the operation of wind electric generators and their interconnection with the grid. And today this focus has culminated in a fully fledged domestic industry, which now sells wind turbines on the world market.

Today the MNES is involved in the development, demonstration and utilization of various renewable energy-based technologies, including: solar thermal; solar photovoltaics; wind power generation; wind-powered water pumping; biomass combustion and co-generation; mini and micro hydropower; solar power; utilization of biomass gasifiers; biogas; improved chulha (cookers); energy from municipal and industrial wastes; and tidal power generation. The MNES also deals with other emerging areas and new technologies, such as new chemical processes, fuel cells, alternative fuel for surface transportation and the use of hydrogen as an energy carrier.

State-sponsored R&D support

The ability to move an indigenous industry to the front of the international field requires that industry has access to scientific and academic expertise. Solar PV is a useful example of an extremely high-tech industry defined by ongoing innovation. India is one of largest existing markets for solar home systems. In India, research and development has been a major component of the Indian Solar Photovoltaic programme. In addition to the establishment of an indigenous manufacturing capability, continuing R&D efforts have led to reductions in costs, improvement in reliability and the introduction of newer technologies. As discussed above, India's PV market has been driven by a long-standing government programme of subsidy, tax and financial incentives for two decades. With the increase in volumes now achieved, there is a shift in policy in favour of commercial, market-oriented approaches as well as technology R&D.

India has also taken steps to develop renewable technologies that have lagged behind leaders like PV and wind, but which hold significant potential for India and many other countries. One example is solar thermal power systems based on solar thermal concentrators. Systems have now been developed to increase the temperature of a working fluid to above 300°C, running a conventional turbine to generate electricity. A 140MW Integrated Solar Combined Cycle (ISCC) Power Project is now being established by the government of India, MNES at Mathania in the Jodhpur district of Rajasthan.

National standards and testing

Another key role for government is to ensure that markets are not disrupted and that consumer confidence in the new products and technologies is steadily built up. One example is the use of national standards to check for product quality and service quality. By way of illustration, Indian solar water heating systems and solar cookers have seen product innovation and cost reduction carried out in order to make the solar thermal devices market oriented. They are now considered to have been fully commercialized in India. To assure quality and industry standards, the products are now standardized by the Bureau of Indian Standards (BIS).

Similarly, in the field of wind energy, the MNES has established a wind turbine test station, under the Centre for Wind Energy Technology (C-WET), to focus on standardization, testing and certification in order to improve performance levels in the manufacturing, installation and operation of wind electric generators in India. The wind turbine test station at Kayathar was dedicated to the nation by the Prime Minister of India in July, 2000. The test certificate for the first 500KW wind turbine tested there was presented by the Honourable Union Minister of State (Independent Charge) for Non-conventional Energy Sources of the government of India, on 13 August 2001 at Chennai, Tamil Nadu.

Prioritized national investment

As we have indicated above, one of the reasons for the growth of the renewable energy industry in India is the conducive investment policy regime pursued by the government of India, and this deserves a little more attention.

The renewable energy sector in India has been accorded 'priority sector' status by the Reserve Bank of India for the purpose of making available loans from banks. As a result of these initiatives, a strong manufacturing base backed by International Industrial Partnerships (IIP) has been created for this sector in India.

In order to accelerate the promotion of renewable energy technologies and systems, the MNES set up a financial institution – the Indian Renewable Energy Development Agency Ltd (IREDA) – as a public limited company in 1987 under the Companies Act, 1956. The main object of IREDA is to provide financial assistance to technologies and projects related to NRSE. Wind energy is one of the largest beneficiaries among the NRSE sectors and accounts for approximately 50 per cent of the total loans disbursed by IREDA.

Integrated national, state and regional policy-making

The role of state and regional governments in ensuring a consistent policy framework has been an important feature of the success of Indian renewable energy policy-making and it is worth considering how this has been achieved.

A central aspect has been communication and inclusion. For example, the MNES issued guidelines in July 1995 to all the state governments on policies conducive to power generation from renewable energy sources with a view to encouraging commercial development in the NRSE sector, including wind energy.

For wind energy, around nine states with wind potential have announced attractive schemes for encouraging the private sector to generate power from harnessing wind energy. MNES also issued guidelines in 1995 to all the private sector developers, to ensure that the incentives and facilities provided by the central and state governments were properly utilized by the developer and also that due emphasis was given to quality in the selection of wind electric generator. The guidelines issued by MNES to the state governments and the wind power developers constituted an integrated government and private sector approach to the harnessing of wind energy in India.

Similarly, an integrated priority approach has been accorded to the promotion of commercial small hydropower projects through the private sector. MNES issued guidelines to all states for adopting a uniform policy in relation to non-conventional energy power projects. Already 13 states have announced their policies to encourage private sector investment in commercial small hydropower projects. In these states, sites with a total potential of over 2000MW have been offered to the private sector for setting up commercial projects in the next 10 years.

Indigenous manufacture and exporting

The reward for leadership and early adoption in any commercial field is the ability to export products and services to those that come later.

The MNES and its financial arm, IREDA have, through their strategic planning and implementation of various programmes, built a solid base for the growth and development of non-conventional energy sources in the country. This stable domestic market is important in itself. However, it also puts industries in a favourable position with regard to overseas markets as those industries reach a sufficient level of maturity. Indian industries have absorbed and adopted technologies in the different disciplines of non-conventional energy sources and are now capable of producing products and systems to internationally recognized standards and quality.

By way of an example of a specific intervention, in order to accelerate indigenization of wind electric generators, the government of India has imposed a customs duty on complete wind electric generators since 1992, but has permitted the import of specified parts and components required for the manufacture of wind electric generators under nil/fully concessional duty. Also, wind electric generators are exempted from the payment of excise duty and sales tax. The clear aim – and consequent result – has been the encouragement of a number of international and Indian wind turbine suppliers to set up indigenous Indian production facilities.

At least ten domestic companies are manufacturing wind electric generators and components, either as joint ventures or as licensees from international collaborators. Collectively they have achieved an annual turnover of 15 billion rupees. Some of these companies are now active in markets beyond Indian shores. Wind electric generators of multi-megawatt unit capacity are now being manufactured indigenously. Even rotor blades, a crucial and high-tech composite component of wind electric generators are now manufactured in India. Nearly 80 per cent indigenization has been achieved with manufacturing to European commercial and insurance standards.

The same process is also underway in other renewable energy technologies. For example, a major solar thermal company set up in India in the late 1980s, based on Canadian technology, and today is exporting its products to Spain, Germany, the Netherlands, Denmark, Greece, Turkey, Cyprus, Korea and Africa.

International Policy Opportunities

With the Kyoto Protocol ratified, there are now multilateral policies starting to emerge that affect energy development. One of first policies and mechanisms, among what may become many, is the Clean Development Mechanism.

India and the Clean Development Mechanism (CDM)

The Clean Development Mechanism (CDM) is set out under the Kyoto Protocol as a cost-effective flexibility measure to mitigate climate change and to promote the transfer of climate-friendly technologies. The CDM is designed to assist developing countries achieve sustainable development, to contribute to the ultimate objective of the United Nations Framework Convention on Climate Change (UNFCCC) and to assist the parties in achieving compliance with their emission limitations.

Under the CDM, countries cooperate in an emissions mitigation project in a developing country with the donor country acquiring the Certified Emission Reduction Units generated by the project, while the host country benefits from the contribution of the project to sustainable economic development through investment in environmentally sound technologies such as renewable energy technologies (RETs).

Also, the Conference of Parties in Marrakesh agreed that, in order to assist small projects to overcome the transaction costs involved in executing a CDM project, to ask the CDM Executive to develop simplified modalities and procedures for small-scale renewable energy project activities with a maximum capacity up to 15MW.

A CDM project under the Kyoto Protocol is a development project, driven by market forces, which reduces greenhouse gases against a validated baseline. The project investor takes a risk based on the returns provided by certified emission reduction units, which correspond to the reduction of greenhouse gases achieved. The CDM also provides a trade opportunity for the developing countries to collaborate with industrialized country investors to develop new industries and technologies. This form of trade in credits under the CDM requires that projects in developing countries must contribute to the sustainable development of the host country. The decision regarding which project does and which project does not contribute to the sustainable development of the host country is left to that host country itself, which must under the Protocol create a Designated National Authority to approve or reject project applications. The trade is registered only if the CDM project results in real and measurable reductions in the emission of greenhouse gases, which is to be validated, monitored and verified.

For a country like India, the CDM offers an opportunity to promote sustainable development and to direct the flow of capital, expertise and technology into developing countries.

The government of India had already constituted a Climate Change Advisory Group on Renewable Energy in order to take advantage of the vast opportunities available to the renewable energy sector under the CDM project and similar schemes anticipated post-2012.

The Future Direction of Indian
Renewable Energy Policies

It is important for national policies in India to keep up with the maturing industry. Already a comprehensive renewable energy policy for all-round development of the sector has been formulated by the MNES. Policy measures aim at overall development and the promotion of renewable energy technologies and applications. Specifically, these will focus on policy initiatives encouraging private investment and Foreign Direct Investment (FDI).

The growth in demand for renewable energy has had profound implications on industry and has also led to a growth in international industrial partnerships in the renewable energy sector. A new thrust is being given to consolidate the progress made so far, by undertaking efforts to remove constraints and bottlenecks for accelerated utilization, and also to venture in new directions across the entire spectrum of rural energy, solar energy and power generation by using RETs.

Although the MNES has been supporting R&D for technology and human resources development in the renewable energy sector, the emphasis now is on cost reductions and increases in efficiency. It is intended that there be increased interaction and close cooperation between the the country's research and teaching institutions, which are reservoirs of knowledge and experience, and Indian industry, which has the requisite entrepreneurial skills and market orientation.

Conclusions

In India a significant thrust has been given to the full chain of renewable industry development from research, development, demonstration and actual implementation of various renewable energy technologies and thence to indigenous manufacture and exporting to international markets.

Nationally, renewable energy is seen as an effective option for meeting ever-increasing energy demands as well as a means of providing sustainable national energy security with important consequent benefits to the population and its current needs and aspirations.

Some of the policy areas we have identified in this chapter include:

- establishing powerful government agencies to develop the sector;
- comprehensive resource assessment to guide the industry;
- target setting by government to send messages to the private and public sectors;
- the identification of target markets and consumers;
- focused market development and technology development;
- utilization of business law to stimulate technology transfer;
- fiscal intervention to create appropriate market dynamics;
- utilization of joint venture laws to accelerate technology transfer;
- R&D support;
- implementation of standards and testing;
- prioritizing national investment;
- integrated policy-making – national, state, regional and international;
- using trade rules to encouraging indigenous manufacture;
- orientating the mature market towards international competition.

There have been successes and failures over the three decades during which India has looked to harness its renewable energy resources. However, the current success speaks for itself in terms of the implemented renewable energy projects and the ability of many Indian companies and technologies to engage in new and emerging renewable energy markets.

The momentum continues to build and it is estimated that in the next 10 years India will be requiring an additional capacity of around 100,000MW, and around 10 per cent of this (10,000MW) is expected to be contributed to the Indian grids from the NRSE sector. India is, and will continue to be, a place of great opportunity for the global and national renewable energy industry.

9

Spanish Renewable Energy: Successes and Untapped Potential

José Luis García Ortega and Emilio Menéndez Pérez

The Spanish Energy Context

Energy dependence and emissions

Spain is highly dependent on imported energy supplies in various forms; three-quarters of all basic energy requirements are imported. Nevertheless effective energy efficiency programmes have not been put in place, even despite public concern about environmental issues such as climate change.

Energy consumption has been growing rapidly, increasing greenhouse gas (GHG) emissions well above levels agreed by Spain and the European Union toward its Kyoto commitments. Emissions levels, permitted to rise by 15 per cent for the 2008–12 Kyoto commitment period, had instead ballooned more than 40 per cent by 2003.

Transport

Car numbers have slightly increased to about 500 per 1000 inhabitants, while there is a shortfall in public transport frameworks both within cities and between them. Domestic and foreign tourism has greatly increased too. Thus, automotive fuel demand is rising, and the combined effect is that Spain's transport GHG emissions are the fastest growing of all sectors.

Electricity

Demand for electricity has also been increasing rapidly, from 135 terawatt hours (TWh) in 1990 to about 250TWh (approximately an 80 per cent rise). Electricity consumption for domestic use and services has risen significantly. The massive uptake in household appliances as well as a growing demand for air-conditioning continually pushes energy demand upward, particularly in major cities and tourist areas. This was starkly demonstrated by the record-breaking electricity demand for June–July 2004, when temperatures reached their highest peak ever in cities including Madrid.

Fuel sources

The country is highly dependent on crude oil; half of Spain's energy use is related to oil, almost all of it imported. The oil-price crisis of the 1970s had a severe impact on the national inflation rate, which rose up to 20 per cent per year with huge economic impacts. During the 1980s and 1990s, coal made a comeback as an energy source to generate electricity and to fuel larger industries such as cement production.

Nuclear energy was also introduced as a supposed solution. The building of nuclear power stations and the regular emissions of radiotoxic waste in the Atlantic marine basin of Galicia incurred a heavy social response. Less than a quarter of the proposed plants were built. Nevertheless, nuclear energy generates almost one third of Spain's electricity.

Spanish gas supply has been the focus of a far-reaching development scheme to secure external supply and internal demand. This development and distribution is intended for domestic and commercial uses as much as for electricity generation. Approximately a third of the gas supply is expected to displace coal-fired power generation. This is expected to have impacts on CO_2 emissions, although less than the equivalent displacement by renewables.

The History of Renewables Development

During the first half of the 20th century, hydropower was the foundation of the Spanish electricity scheme. Although its use began with plants of low capacity, during the period 1940–60 larger installations were built. As in many countries, the scale of these installations became the source of a major social clash of interests, particularly in the states of Galicia and Aragon.

Today hydroelectricity contributes 15 per cent of the country's electricity production, subject to annual rainfall. However, it plays a greater part in terms of installed power capacity, at 16GW or one third of the total 58GW. This underscores its continued role as a peak demand supply despite the comparatively low capacity factor (average power output as a fraction of maximum rated power output).

In the second half of the 1970s, following the oil shock, the Spanish government took part in international initiatives to enhance renewable energy development. An example was the Plataforma Solar de Almería (Almeria Solar Platform), established with joint Spanish–German funding. Three different small, solar thermal electric power stations were built and the R&D work undertaken here has continued through two decades.

The above-mentioned oil crisis also gave birth to a new national agency, the Centro de Estudios de la Energía (Energy Research Centre). It later became the

Instituto de Diversificación y Ahorro Energético (IDAE; Energy Diversification and Conservation Institute). The IDAE played a significant role in the development of renewable energies and especially policy programmes, as we shall see.

Until the early 1990s the focus was very much on technical innovation and basic research and development. Meanwhile, as social interest in renewable energy gathered pace, some smaller companies became established while larger companies saw the brand advantages of being associated with clean energy.

At the same time, three key pressures were building support for renewable energy promotion. These were the increasing dependence on external energy sources, the prospects for job creation from new industries and technologies, and increasing environmental awareness.

The renewable energy advocates

In Spain two key groups have emerged as the most active and vocal advocates for renewable energy and the systemic changes needed to bring about its development: independent developers and civil society organizations.

The independent developers' role was a key one. Smaller than traditional utilities, they also differed in that they wanted to use renewable energy – mainly small-scale hydro and wind. To protect and promote their common interest, these firms formed umbrella associations. This allowed them to speak with a unified voice and the content, in many cases, was quite different from what upper management of traditional utilities were saying about renewable energy.

We must also emphasize society's role and that of environmental groups in particular. For the major nationwide environment groups, renewable energies became the chosen alternative for tackling energy-based pollution. Such technologies allowed these groups to reconcile protection of environmental assets with a global and integrated solution and perspective. Importantly, the support and political pressure exerted by environmental and civil society groups in Spain preceded the development of the domestic industry.

A 1992 agreement to promote wind power provides an example of the pressure that multi-stakeholder coalitions applied to foster Spanish renewable energy development. At a time when the official IDAE objective was 175MW, a coalition comprising a nationwide environmental organization (Asociacion Ecologista de Defensa de la Naturaleza (AEDENAT)), along with two major national unions (Union General de Trabajadores (UGT) and Comisiores Obreras (CCOO)) and the Empresa Nacional de Electricidad SA (a publicly owned national electricity company at that stage) came together and called for a 750MW target. Although the target appeared overly ambitious at the time, their aims were to turn Spain into a leading country in a new environmental field, secure jobs and develop domestic energy security. Most would now consider these objectives achieved, or at least well on the way.

Once the renewable industry became more established in the 1990s, the role of pressure groups evolved. They now fill the following roles: leveraging required increases for renewable programmes and projects, breaking political, policy or systemic deadlocks that emerge, and taking the lead to counter social lobbies against renewable energy.

New incentives: The royal decrees

Although Spain is now recognized internationally as a wind energy leader, as recently as 1993 there were only 52MW of wind power operating. However, the following year new policies had dramatic effects. Legislation enacted in 1994 obliged all electricity companies to provide a five-year minimum price for energy generated from renewable sources.

This policy reform took place in a climate of extensive energy reform in the wider power sector and a gradual transition to a liberalized market. A few days before the electricity sector liberalization act was approved in 1998, the government consolidated the role of the renewable energy sector with a Royal Decree that harmonized the incentive regime within the emerging competitive electricity markets. Furthermore, the 1998 law set down an objective of Spain sourcing 12 per cent of its energy from renewables by 2010.

Structure of support

Renewable energy promotion is based on premium payments for renewable energy generation. This renewable bonus varies according to renewable energy source or technology and is generally updated annually by the government and subject to a four-year review. For example, during the 2002 period the allocation was: wind, €0.029 (US$0.039); small-scale hydro, €0.031 (US$0.042); biomass, €0.0279 (US$0.0376); solar thermoelectric, €0.1202[1] (US$0.1618); photovoltaic up to 5kW, €0.3606 (US$0.4853), and photovoltaic of more than 5kW, €0.1803 (US$0.2427).

A new Royal Decree was enacted in 2004 with two main targets. The first was to create incentives and mechanisms for renewable power to participate in the liberalized electricity market. The second was to provide a longer-term existence for the premium payments, remaining untouched at least until the respective capacity objectives are met, thus guaranteeing confidence for investors.

Support for solar PV power was enhanced by raising the size limit of projects eligible for higher premiums to 100kW. This led to soaring demand for solar PV (from a rate of 4MW being installed over an entire year to petitions of 40MW in just a few months, with 10MW installed during its first year). Solar thermoelectric power has also had its premium payment level raised to the equivalent of €0.18 (US$0.24) per kilowatt hour, thus making new projects feasible.

Despite the new decree, there are still concerns that small independent developers feel have been unaddressed. Thus they have called on the new government, elected in 2004, to remove aspects that penalize small producers, and also to create improved incentives for biomass.

The Role of the Spanish States

Spain is a country geographically divided into different regions known as comunidades autónomas, each of them with their own parliaments and governments. The regions also have agencies promoting the efficient use of energy and renewable energies. Some of these state agencies have had a very significant role in renewable energy development. Furthermore, many different town and county councils have established their own energy agencies aiming to locally promote renewable sources of energy.

With regional unemployment figures that can reach as high as 10–20 per cent, a central issue for states has been job creation. Many have chosen to link state renewable project development approval with investment that ensures money flows into the local economy.

Furthermore, the states have undertaken aggressive targets to push activity along. For example, Galicia, which has an Atlantic coastline, has had a 2300MW wind target since 1997 – equivalent to 45 per cent of the state's power demand. This has been tied to the aim of ensuring that 70 per cent of the investment is spent inside state borders. It has resulted in over 5000 direct and indirect jobs and numerous factories.

Another central role for the states has been to engage on environmental issues. The state of Navarra included environmental issues as part of the key parameters in site identification at the outset of planning. This prevented some of the conflicts that subsequently held back development in other states.

The role of municipalities has also become significant in issues such as solar thermal promotion. One success story is an initiative by the Barcelona council called the 'solar ordinance'. This by-law requires all newly-built and refurbished buildings in the city to be equipped with solar thermal collectors capable of supplying at least 60 per cent of their hot water demand. After one year of effective

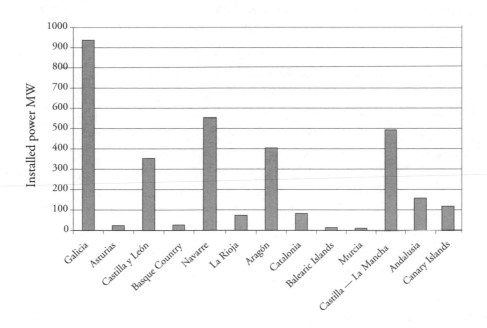

Figure 9.1 *Installed renewable energy capacity in Spain's various* comunidades autónomas

implementation of the solar by-law, the area of solar thermal collectors in Barcelona quadrupled. Several cities including Madrid, Seville, Granada and Pamplona have replicated the initiative. Now 20 per cent of the Spanish population lives in municipalities that require solar thermal for hot water in new buildings, and this approach is being considered by some states. However, it is still too early to measure the market impact of these initiatives.

Outcomes: Installed Capacity and Energy Generation

Obviously a key measure of success is the quantity of renewable energy actually produced in Spain. Table 9.1 and Figure 9.2 show the progress of some of the major renewables in Spain (the absence of solar thermal electric power from Table 9.1 will be explained later in the 'Technology choices' section).

Some Persistent Resistence

So far in this chapter we have focused on the successes of renewable energy in Spain and the path of its evolution. However, this path has not been without obstacles and it is definitely worth providing an overview to show where the main challenges have emerged.

Table 9.1 *The evolution of various types of renewable electricity sold to the grid on the Spanish mainland in the period 1990–2002*

Year	GWh			
	Wind	Small hydro	Biomass	Solar PV
1990	2	977		
1991	3	1647	1	
1992	17	2037	5	
1993	85	2241	14	
1994	73	2491	55	1
1995	160	2240	203	1
1996	304	3589	235	1
1997	595	3451	107	1
1998	1237	3618	165	1
1999	2474	3786	188	1
2000	4462	3914	267	1
2001	6600	4385	628	2
2002	9220	3893	1084	4

Source: Comisión Nacional de la Energía (National Energy Commission)

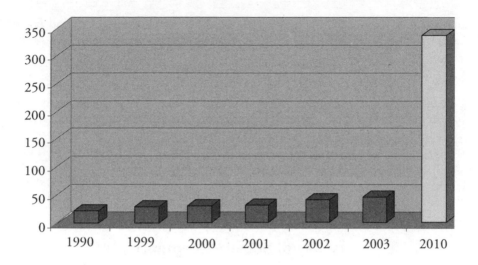

Source: IDAE, 2004

Figure 9.2 *Evolution in demand for solar thermal energy (low temperature) in kToe (thousand tons of oil equivalent) compared with existing government target*

Industry

In the early years, companies from the Spanish electricity system participated in the promotion of renewable energies. From 1980 on to the mid-1990s a Programa de Investigación Electrotécnica (Electrical Engineering Research Programme) was funded using 0.3 per cent of the electricity charges. It also used other funding from the European Commission. Despite the programme being well positioned to become a major national driver for renewable energy development, investment priorities have and continue to be assigned to traditional generators. This has tended to mean that some of the main renewable energy proponents are third-party players sitting outside the existing power industry. This has led to an obstructive, competitive dynamic in which utilities have been accused of using grid access as a major obstacle to renewable energy developments.

Government

Some of the more critical elements of policy-making can occur when a renewable energy technology expands sufficiently to be perceived as capable of direct competition with conventional energies; this currently appears likely with wind power. Bending to pressure from some of the larger traditional utilities, government has placed caps on the total amount of wind energy permitted into the grid. Such policy-making – included as a last-minute annex in the Energy Infrastructure Plan – represents a serious threat to long-term domestic investment in renewable industries. However, the newly (post-2004) elected government is negotiating with the grid transmission operating company to agree amendments that could let wind power capacity reach 20,000MW (installed capacity was 8263MW at the end of 2004).

Society

Like other countries, Spain has seen some social groups emerge to oppose renewable energy projects or technologies. This may arise through a lack of understanding of the overall issues, because a group has a smaller geographic focus, or sometimes because a group is focusing on a very specific aspect of the environment and is ignoring the wider consequences of the energy system. Whatever the reason, these groups can end up being perceived as the voice of wider society unless the mainstream environmental sector maintains high-profile support for renewables.

Issues of Social Acceptance

Manufacturing and job creation

One reason for the strong social acceptance of Spanish renewable energies has been their capacity to create jobs, even more so when or where unemployment is a serious problem. Figure 9.3 depicts direct employment and future projections. As the figure shows, there were nearly 20,000 jobs by 2001, and by the end of the decade expectations are that approximately 45,000 people will be directly employed in these industries.

Of the renewables technologies, the wind industry generates the most jobs in Spain with approximately 10,000 jobs overall. Blade manufacture alone creates 1000 direct jobs in seven factories, and 1000 more jobs are created by small operation and maintenance enterprises all over the country. Offshore wind development's employment potential would have special resonance for shipyards that are going through industrial restructuring and consequent job losses.

Solar power is believed to employ approximately 4000 people in Spain, with three-quarters of them in small operation and maintenance enterprises. PV module manufacturing requires more than 1000 workers and 15 per cent of them are graduates of higher education.

Employment in biomass generation takes the equivalent of 5000 full-time people; most of this labour is in fact complementary to work already done collecting agricultural waste in rural areas.

Financial impacts

Regional governments throughout Spain regulate the installation of renewable energy plants in their own territory, for which they have several planning approaches. For example, wind developers are required to pay landowners (whether private or public owners) a rent which constitutes 1.5 per cent of the total value of the power generated.

However, it has also been important to ensure that the wider community benefits from renewable energy development. This has occurred through the pay-

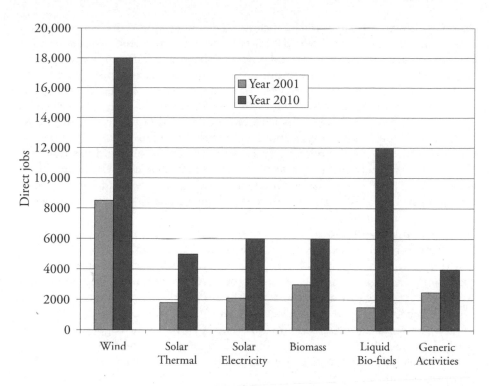

Figure 9.3 *Estimated evolution of direct employment
in Spain by renewable energy segment*

ment of local taxes and especially through the creation of local jobs – as discussed above.

Environmental impacts

Although the environmental repercussions of wind power plants have been moderate, the expansion of wind power in Spain has not been without incident.

A very high-profile case in which vultures died through impacts with wind turbines was due to locating the wind farm close to a refuse dump where the birds scavenged. Changing the dump's location solved the problem. However, it would have been far better if such impacts had been avoided altogether since these stories have a long life. Other consequences, such as land erosion, have been raised but have turned out to be of little real concern. What has instead been seen to be crucial are the impacts on the human environment, not just for wind, but also for small hydro and biomass.

Social impacts

With many of wind power's environmental issues already having been addressed, it is the issue of aesthetics that more often takes the stage, primarily in relation to the location of generators on the top of mountains in Spain. Throwaway comments on the radio, for example, might include, 'Let's hope not many wind turbines appear on our way when we take the car for a drive this weekend.' Aesthetics may become a crucial issue to future wind power development in Spain. Development is now taking place in diverse locations, including rural areas where there is low economic activity. This can be seen as a welcome source of external investment in such areas, but it can also be cast as selling these areas out to external interests.

Another feature of Spain is its Mediterranean location which contributes to high levels of tourism. Concerns about wind turbines' effect on tourism must be put into context with the ongoing and severe impact of tall buildings built for the tourism sector. In practice, however, this sort of opposition has largely been generated by local groups that are focused on issues of poor planning and management on the part of companies or regional governments, rather than on wind energy development itself.

As for small hydro the public reaction has not been inspiring. Public and private agencies have been examining the possibility of re-commissioning abandoned hydro sites from the first half of the 20th century as well as looking into new sites that combine irrigation and water supply services. Nevertheless, the public still appears to have negative associations with hydro development even though some earlier projects were quite sound, such as that in the Pyrenees.

The current location of small hydro-power plants, on the edges of rivers and small dams are places where farmers, stockbreeders and fishermen often go for work and where many others go for recreation. Few are prepared to support further impacts on these areas.

Unfortunately, the economic benefits of hydro are seen to be comparatively small and it is not seen as a defining technology or industry of the future. In 2002, the installed capacity was 1492MW[2] – less than 2 per cent of the energy used in Spain. However, a comparatively modest increase to 2200MW is planned by 2010.

Room for Improvement

Although there is a sense in Spain that the bonus approach to funding renewable power generation has largely been effective, the runaway success of the wind industry should not blind us to the reality faced by other industries. There are areas of policy weakness that can be identified when the various technologies are considered.

Integrated planning

We have seen that several types of organizations and agencies were established by state and national governments to assist renewables. However, the Spanish government policy on energy has not always been completely helpful. Generally speaking, priority has focused on traditional ways to generate energy rather than on renewable ones.

Consider the following two plans, for example. The Plan de Ahorro y Eficiencia Energética (Energy Conservation and Efficiency Plan) which is attached to Plan Energético Nacional 1991–2000 (the 1991–2000 National Energy Plan), and the

1999 Plan de Fomento de las Energías Renovables (Renewable Energies Development Plan) prior to Plan de Infraestructuras Energéticas 2002–11 (2002–11 Energy Infrastructures Plan). One of these plans promotes renewable energy while the other promotes conventional sources as the basis of Spain's energy system.

This sort of energy planning discrepancy, in which both renewables and conventional energy are promoted, can be confusing. This is true for long-term rather than short-term energy planning. Even today, true integration of renewable energy sources and energy efficiency into energy system planning is lacking, as is integrated resource planning which treats all energy options equally.

Taxation

Taxation of fuels is the dominant determinant in pricing, uptake and investment. Petrol in Spain might typically cost €0.90 (US$1.21) per litre, two-thirds of which is tax. By comparison 1 litre of bio-alcohol costs about €0.75 (US$1.01) to produce, including paying the farmer for the raw material and farm labour in sufficient quantity to overcome EU payments for leaving the land fallow.

Bio-fuels are identified as one of Spain's priority energy options with a proposal for unused agricultural land to be turned over to bio-fuel production. Spain's Plan de Fomento de las Energías Renovables (Renewable Energies Development Plan) sets a goal of 500,000 tonnes of oil equivalent by 2010 which is nearly 2 per cent of the fuel utilized by Spanish cars.[3]

The central issue here is to determine the appropriate taxation level. The Spanish administration recently decided to de-tax liquid bio-fuels, thus enhancing an emerging industry of recycled vegetable oils for the production of bio-fuels. However, the balance of costs is still an economically unfavourable option for farmers.

Access

We mentioned earlier the problem of utility control over third-party grid access as a potential obstruction to developments as well as a cause of cost increases through delays.

A similar aspect can be seen in bio-fuels. Because control over distribution is in the hands of oil companies it is therefore desirable that these companies permit ethanol to enter their distribution structure (as an additive or higher percentage mix) and made available at petrol stations.

Many commentators believe a legal obligation is necessary to ensure bio-fuels' use in the fuel market, perhaps combined with regulations to impose efficiency standards for vehicles.

Scale and variability

Many renewable technologies entail the development of non-standard projects. This has proven to be a hurdle that the simple bonus price support scheme has not adequately addressed.

For example, power generation from biomass is progressing slowly. Thus far, three power plants (totalling almost 50MW) have been built using polluting olive industry wastes, and an installation (25MW) powered by cereal straw has also been built. Yet the industry will not only fall well short of the Development Plan target of 1700MW in 2010, it is quite likely to achieve only a mere one third of this target.

There are several reasons for this. First, investment in power plants is too high to be undertaken by rural communities or companies, even though they may be the best placed to identify the opportunity. Second, utilities may tend to avoid such projects as they involve time-consuming negotiations with a new set of farmers for each project. Third, utilities tend to be drawn to relatively large projects with better economic potential – such as concentrated crop wastes – leaving smaller-scale projects unrealized. And finally, the current fixed bonus system is not suited for an industry as heterogeneous as biomass, the costs of which vary with the different fuels and technologies. It would be useful to further diversify the bonus (to provide the varied projects with the incentive they require to develop), taking into account technical, economic and environmental parameters.

Technology choices

Spain has successfully demonstrated the transfer of technology from lead countries and become a significant international player in renewables. However, there are areas where Spain should have led from the start but did not.

The development of solar thermal electricity in Spain provides an excellent demonstration of the importance of making careful decisions about technology and resources, and not just using received wisdom.

As mentioned earlier in this chapter, at the end of the 1970s the Plataforma Solar de Almería was created as a technology R&D centre and three different, small solar power stations were built, with mirror modules that concentrated solar radiation to levels sufficient to generate electricity in a thermodynamic cycle.

The recent bonus approval, equivalent to €0.18 (US$0.24) per kilowatt hour mentioned previously makes proposals for four more 10–50MW solar thermal electric plants possible; three in Andalusia and one in Navarre. They have mirror and tower solar fields, cylinder-parabolic trough collectors, and one of them incorporates a new thermal-cycle design using air as an intermediate fluid for high-temperature heating in tower boilers.

However, what is less than ideal is that the bonus scheme for this type of technology was not introduced alongside other renewables at the outset in 1994/98, but had to wait four years. Despite this oversight, Spain, with its intense solar resource and appropriate policy framework, is now a world leader in this type of technology. Under the new Royal Decree expectations from industry are that a new 1000MW target for this technology looks more realistic than the former government's Renewable Energies Development Plan target of 200MW by 2010.

Research and development

One of Spain's main R&D institutes is the Centro de Investigaciones Energéticas, Medioambientales y Tecnológicas (Energy, Environment, and Technology Research Centre), where 200 people are employed.

The centre's various activities include: wind technology development, both for big on-grid machines and small off-grid installations; new processes to obtain liquid bio-fuels from large volumes of raw material; thin-film PV technologies and development of new materials; and solar thermal electric research and demonstration projects.

However, in practice, there appears to be a gap between this research and the world of industry. Spanish companies continue to seek technology from outside the country even as our researchers turn to European Union projects to raise funds to show they can provide a useful contribution. This indicates that the Spanish administration and the technology centres have not managed to integrate enterprise building and national technology research in energy.

The Solar Energy Institute is an example of what can be achieved when research is closely integrated into the nation's industrial development needs for renewable energy. The Spanish market only absorbs about 4 per cent of the modules made in Europe, yet successful industry, research and policy integration has resulted in more than 95 per cent of Spanish PV module output being exported. This includes exports to the rest of the EU, Latin America, Africa and Asia. Spain is now a leading European PV manufacturer and has demonstrated what can be achieved when things are done right.

Notes

1   The government in July 2002 approved this figure. Up until then, the bonus for this technology was only €0.03 (US$0.04), causing a delay in its commercial development.
2   This is according to Comision Nacional de Energía (CNE) information on statistics on energy purchase to the special regime (CNE, 2003).
3   In 2005 a new Renewable Energy Plan was approved by the Spanish government, raising the target for biofuels to 2,200,000 tonnes of oil equivalent by 2010.

References

CNE (2003) *Informe sobre las compras de energía a régimen especial. Period: Año 2002*, Comision Nacional de Energía
IDAE (2004) *Boletín IDAE: Eficiencia Energética y Energías Renovables (No 6)*, Instituto para la Diversificacion Ahorro de la Energia

10
A History of Support for Solar Photovoltaics in Germany

Sven Teske and Volker U. Hoffmann

Introduction

Along with Japan and the US, Germany is now recognized as a country at the forefront of the drive for photovoltaic (PV) industry development. However, the journey to this current status and the supporting policies that guided it involved a rocky path with many lessons being learned along the way.

This chapter outlines four key stages in the life of German solar power since the early 1990s, from the inception of the Thousand Roofs programme, through the years of non-integrated regional support, and onto the 100,000 Roofs programme. Finally, it will describe the evolution of the Renewable Energy Law towards inclusion of a pure feed-in tariff scheme under this law.

1990–95: The Thousand Solar Roofs Years

Germany has a long history of scientific research and technical innovation; the energy sector has seen its share of government financing in this regard. However, when one reviews the spending on innovative energy research between 1974 and 1995 it is quite apparent that the assumption was that nuclear power would be the energy source of the 21st century.

Solar PV has undergone explosive growth in recent years while nuclear fusion has failed to deliver anything approaching commercial power applications. Thus this spending emphasis appears to have been a poor predictor of the energy systems that would prove successful. However, one must remember that even at the end of the 1980s there were still only 15,000 solar installations on homes in the entire world (Fraunhofer Institut, 1997) whereas the nuclear industry had risen to provide significant global generation capacity in only a couple of decades.

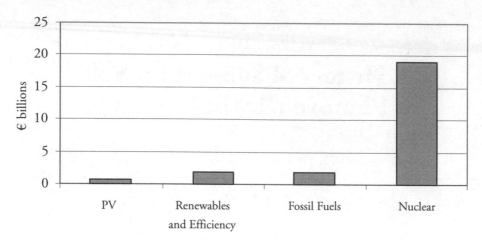

Source: Fraunhofer Institut (1997)

Figure 10.1 *German Research Ministry spending on energy from 1974 to 1995*

The new impetus

The first major solar market creation programme was initiated by the German government in 1991. The German Ministry of Research began testing a wide range of installations connected to the grid in its Thousand Roofs scheme, a joint federal–state programme of support for PV production of power. Its aim was to assess the state of the art already attained by PV technology and from that deduce the development still needed for small-output, grid-connected PV installations.

The name 'Thousand Roofs' caught on as the project title and was retained well after the actual number of installations more than doubled this target.

Structure

The permitted rated capacities for the solar generators under the Thousand Roof scheme ranged from 1 to 5 kilowatts. The total subsidies amounted to 70 per cent of investment costs, with an upper limit for specific investment costs set at DM27,000 per kilowatt peak (kW_p; US\$18.061 per kW_p).

Access to the scheme was by application. At the applications phase of the Thousand Roofs programme there were approximately 60,000 general enquiries which resulted in some 4000 actual applications.

The state allocation

At the inauguration of the Thousand Roofs scheme, each German state received a contingency for the maximum installations. City states had a contingency for 100 installations and other states were allocated 150 (KFW, 2000).

The Thousand Roofs programme was originally conceived for the then-West German states. When full German reunification took place in 1990, the programme was extended to the East German states. Now with 16 states, the maximum number of installations possible under the programme was raised to 2250.

When it became clear at the outset that the states of Mecklenburg-West Pomerania, Saarland and Saxony-Anhalt would not exhaust their contingencies, contingencies were reallocated. Baden-Wurttemberg, Bavaria and Lower Saxony thus increasing the number of installations made. There was no further reallocation.

At the close of the installation phase at the end of 1995, a total of 2100 installations were in operation.

Outcomes

Installations

By the end of 1995, the programme's initial and updated arrangements resulted in more than 2100 grid-connected solar PV systems being installed and trialled on the roofs of houses in Germany. This translated into approximately 7.5MW of peak capacity being installed as a result of the scheme, although the overlap with local incentives makes it difficult to be exact about these numbers.

Proof of concept

The programme proved that the roofs of normal private houses are well suited to the decentralized production of PV power. Yet the surfaces of buildings and roofs used in the Thousand Roofs programme still left significant room for better architectural integration of solar systems.

Awareness

The programme contributed to a marked growth in public knowledge about photovoltaics. Public acceptance of the technology significantly increased and this primed the market for the considerable uptake that followed.

Market growth

In the solar PV sector about 11.5MW$_p$ were installed in 1997 alone. It can be assumed that this development was also due to the success of the Thousand Roofs programme and the extensive publicity given to its results.

Cost reductions

The Thousand Roofs programme led to a reduction in the specific costs of these installations. Specific costs of facilities during the time of the programme were usually about DM24,000 per kW_p, or US$16,000 (although they ranged from DM15,000 [US$10,000] to DM36,000 [US$24,000] per kW_p). By 1997, two years after the end of the Thousand Roofs programme's installation phase, the specific investment cost for grid-connected PV installations had declined to DM17,000 (US$11,000) per kW_p.

System performance

The performance ratio defines the relationship between the target yield and the actual yield.[1] The average performance ratio of the PV systems in the Thousand Roofs programme was about 65 per cent. However, it is evident that a performance ratio of 80 per cent is certainly attainable with such small, decentralized installations on the roofs of houses. This figure should be used as a sort of guideline for future installations.

Energy yield

During the period investigated, the average energy yield per annum from an installation was about 700kWh/kW_p. However, the programme revealed that by selecting the optimum components, yields of more than 800kWh/kW_p could be attained – quite significant at a 15 per cent advantage. This shows that the planning and design of small decentralized PV installations is quite important.

Domestic power source

The average output of solar generators in the Thousand Roofs programme was 2.64kW_p. This output, looked at summarily, could on average cover about 50 per cent of the annual electricity needs of the households involved. However, it is possible to install systems of up to 5kW; furthermore consumer behaviour reveals an increased awareness of energy efficiency. This means that it is, in principle, possible for a grid-linked, PV rooftop installation to cover the total annual electricity needs of buildings which house one or two families.

Industry growth

As mentioned above, the programme oversaw the installation of more than 2100 systems with a total capacity of 11.5MW per annum by 1997. Thus the Thousand Roofs programme saw an embryonic PV industry established in Germany. It witnessed a new breed of companies specializing in the planning and installing of household-scale solar facilities. It furthermore allowed existing capacities in making PV components (modules and inverters) to be consolidated or expanded. A new emphasis was also brought to bear on the design and manufacture of specialized inverters.

Room for improvement

Apart from the technical points raised above, some directions for improved policy-making became clear.

One issue was maximizing yield. The subsidies were based on the capacity of the PV systems, but research showed that the average system had considerable room to increase power output. More sophisticated subsidy schemes might be designed which clearly relate to the actual output of installations. The operators of the installations would thus be motivated to ensure the systems worked without disruption and with a maximum energy yield.

A second issue was maximizing uptake. Despite the altogether favourable results of the Thousand Roofs programme and the encouraging public response to it, about half of the applications made could not be approved because of the cap placed on the scheme. Thus an important portion of market demand was left unexploited – and public appetite for PV systems left unsatiated.

Continuity

Contrary to the recommendation of the Commission of Inquiry on Climate, the German government did not set up a successor programme when Thousand Roofs came to an end in 1995. Thus the modest beginning in producing PV facilities was effectively stopped in its tracks in the mid-1990s. This had serious industrial consequences. The two big manufacturers, Siemens Solar (today Shell Solar) and DASA/ASE (today RWE Schott Solar) announced they would cease making cells and modules in Germany and decided to move production to the US.

1995–99: The Dark Ages

The German federal government effectively stopped subsidizing photovoltaics for almost five years after the Thousand Roofs subsidy was spent; the only federal initiative that continued was a solar programme for schools. In this section we review the effects of this lack of support.

Support in disarray

What remained after the cessation of Thousand Roofs were programmes undertaken by German states. They were poorly equipped to take the place of the federal programme, with the result that the industry was left with a labyrinth of barriers.

First was the problem of funding through small state-based budgets. There was also no agreement between the federal government and German states on a unified long-term plan, nor was there any straightforward and un-bureaucratic support. Conditions for support varied in each of the 16 states; in some states there was no support whatsoever for years.

In the northern state of Lower Saxony, to cite one negative example, a long-pronounced subsidy programme came into effect but lasted only weeks. It started in April 1994 and a month later was halted by a budget freeze which lasted until November. In February 1995 it was dispensed with altogether. The scheme received eight times as many applications as it was able to approve in an entire year.

Only in the state of North Rhine-Westphalia has there been an uninterrupted investment subsidy programme between 1995 and 1999. This played a major part in preventing the German solar market from collapsing altogether.

So instead of purposeful support for a technology of the future, there was a bureaucratic maze of programmes without a clear rationale.

Green power

Another important factor in the development of the PV market between 1995 and 1999 was solar subsidy programmes offered by the various electricity suppliers. Green tariffs and solar discounts sought customers who would voluntarily pay more for their electricity or contribute to investment in solar plants.

For example, the Bayernwerk electricity supply company (today part of E.ON) had a 1997 programme called Sonne in der Schule (Sun at School), later followed by a Sonne im Rathaus (Sun at the Town Hall) programme. The Sonne in der Schule programme alone oversaw the installation of more than 500 solar systems and a total output of 660kW$_p$ at schools in Germany's southernmost state.

PreussenElektra (today also E.ON) started a similar programme (SONNE online) in 1998 which oversaw a total of 450 installations with a total output of 500kW$_p$. Finally, the electricity supplier in North Rhine-Westphalia, RWE, employed an environmental tariff which allowed it to install photovoltaic facilities with a total output of about one MW$_p$.

Policy mixtures

The confusion of all the regional and national subsidies may have bewildered many potential customers interested in solar power. Besides varying feed-in tariffs and one-off investment subsidies there were also loans available for solar installations at favourable interest rates. In some cases a combination of all three mechanisms could be applied for, these being:

- *Investment subsidies*: here either a fixed sum or a percentage to a maximum of 49 per cent of the price of an installation (per kilowatt output) was paid.
- *Feed-in payment*: a number of different tariffs, namely higher payments for solar power, have been made by local authorities. These have varied between 20 Euro cents (23 US cents) and €1 (US$1.34) per kWh over a period of 10 to a maximum of 20 years. Lower levels of remunerations were often combined with investment or loan programmes.
- *Loan programme*: loans at favourable rates of interest were the least widespread and also least effective method of subsidy. Experience showed that these – taken on their own – had hardly any positive impact on the PV market.

Despite the confusion of all these regional and nationwide subsidies, there was still continual growth in photovoltaics, albeit at a low level. It was therefore possible to prevent the German PV market from collapsing and manufacturers from going abroad.

The technical components of grid-linked PV installations were continually developed. Reliability, particularly of inverters, improved markedly and installation companies' service networks continued to become more and more professional.

Pressure for new support

Despite the lack of a clear national programme, the market for PV still tripled in the two years following the end of Thousand Roofs. The maintenance and strengthening of such levels of support fell to civil society groups.

After the national programme ended, Greenpeace Germany adopted a campaign to maintain public interest in solar power by employing a broad-based solar information[2] and outreach campaign. The effect was continued high numbers of applications for PV subsidy programmes, which maintained sufficient pressure to prevent regional solar power subsidies from being further dismantled.

Greenpeace Germany also commissioned polls in order to demonstrate the level of support for solar energy – tactics clearly aimed at winning political support for the technology (see Table 10.1).

During this time, regional organizations supporting solar power, including Greenpeace groups in over 40 cities and local authority areas, had been fighting successfully for remuneration adequate to properly cover the costs of solar installations.

In parallel, the new PV manufacturers based in Germany prepared to make a comeback. The overall concept proposed was that solar power generators should have a guarantee that they feed into the grid and be paid at rates covering their system costs. Support here was to be based on actual energy produced and stored in the power grid, rather than being based on the capacity of the solar generator.

Table 10.1 *1996 Greenpeace poll indicates members of the public interested in buying solar systems favour a combination of feed-in payment and investment subsidy*

Subsidy model	Percentage of respondents in favour
Mixture of investment subsidy and feed-in payment	52
Feed-in payment covering costs of $1/kWh	39
Feed-in payment of 50 US cents/kWh	10
Investment subsidy of $2500–$3000/kW	11
Tax relief	12
Green tariff	5
No subsidy	0.5

Source: Greenpeace

Depending on the region, the payment proposed was between 50 US cents and $1 per kilowatt hour for a running period of 20 years. In about 40 (mostly small) cities and local authority areas, it was estimated that 7MW of photovoltaics could be installed. The public message of the campaigns was mainly focused on job creation and the high-tech image of solar PV.

The Age of Enlightenment: 100,000 Solar Roofs and the Renewable Energy Law

After the dark ages of programme disarray, which effectively amounted to a huge, five-year field trial for numerous subsidy mechanisms across the German states, a new start at improved PV subsidies was made possible at the end of the 1990s after a change in government.

100,000 solar roofs

In November, 1999 the 100,000 Roofs programme began. This programme was originally supposed to be a continuation of the Thousand Roofs programme. However, an investment subsidy on the scale required – for 350–400MW worth of installation – was too large to be achieved politically. As a compromise, a no-interest loan for 100 per cent of the total investment was guaranteed.

The effect, however, was disappointing. The boom witnessed a decade before did not materialize.

The Renewable Energy Law

Unlike solar PV, for most of the 1990s wind, biomass and some other large-scale renewable technologies had been operating under a completely different system of guaranteed payment for production. This had turned Germany into a world-leading generator and manufacturer of wind energy.

When the provisions for clean generation were reviewed, policy-makers decided to include solar production in this portfolio. Brought into force on 1 April 2000, The Renewable Energy Law (REL) provided for a payment of 50 Euro cents (67 US cents) per kWh for energy generated from PV panels.

The cocktail effect

Critically, the REL was permitted for use in combination with the loan programme. The resultant combination of the 100,000 Roofs programme and the REL attracted larger PV installations.

Furthermore, in addition to householders, commercial enterprises were permitted to access these schemes. The combination of schemes made solar installations commercially viable propositions in their own right, meaning that the investment community was stimulated as well as the environmentally motivated consumer.

An incredible wave of applications for the subsidies followed. The authorities responsible were at times hopelessly overwhelmed. Time and again the pro-

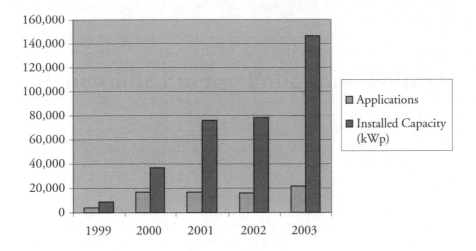

Source: Sven Teske and Volker U. Hoffmann

Figure 10.2 *100,000 Roofs Programme/REL Installations*

gramme had to be stopped, and conditions for loans under the 100,000 Roofs programme continually altered.

Regardless, 2001 was a defining year to date in the PV sector. By 30 April 2002 some 34,000 permits had been given for facilities as part of the programme, and approximately 135MW$_p$ of output installed. And this development has continued on, as can be seen in Figure 10.2.

The peak capacity of the average PV system at the beginning of the 100,000 Roofs programme, 2.5kW$_p$, rose to around 5kW$_p$ in the middle of 2000. Also, increasingly larger systems of up to 100kW$_p$ were built. This appears to confirm the premise that the combination of the 100,000 Roofs programme and the REL attracted purely commercially oriented interests as well.

In June 2003, the 100,000 Roofs programme finished as the overall specified volume had been reached and the government provided no further budget to extend the programme. However, the REL has continued.

REL takes over German PV market

With the 100,000 Roofs programme's termination, the PV market almost fell into a hole. The problem was that the feed-in tariff of 49 Euro cents (67 US cents) per kWh under the REL was not sufficiently high for economic operation of solar PV and the market volume declined. The only reason why the abrupt price support gulf was not seen to be immediately damaging to the industry was that there were still ample PV installations under preparation to satisfy demand for the first half of 2003.

During this time, the entire Renewable Energy Law was under evaluation, which provided an opportunity to address the problem. However, because it soon became clear that this review process would take up to a year, the solar proponents decided that the feed-in tariff for PV should be amended with an extra law.

The joint efforts of the solar industry, parts of the Red/Green government, and environmental NGOs succeeding in pushing through a new feed-in tariff law which came into force on 1 January 2004. The new Renewable Energy Law guarantees that 'the basic rate for electricity produced from solar energy will be 45.7 Euro cents per kWh [61.3 US cents]. If the installations are attached to or built on top of a building, the fees will be 57.4 Euro cents per kWh [77.1 US cents] for an installed capacity of up to 30kW, 54.6 Euro cents per kWh [73.5 US cents] for installations of 30kW or over and 54.0 Euro cents per kWh [72.5 US cents] for an installed capacity of over 100kW.'

The remunerations were also adjusted to the completed 100,000 Roofs programme. For facade-mounted installations, the remunerations were increased by an additional 5 Euro cents per kWh (approximately 7 US cents). The payment of fees for solar electricity is fixed for a period of 20 years.

For PV installations not attached to or built on top of a building structure, fees are only claimable where these installations were commissioned within the scope of a local development plan. This is to ensure that installations do not take place on environmentally sensitive land and that maximum acceptance can be achieved among the local population through community involvement.

It is important to note that the law also forces a steady but manageable price decline on the support tariffs. This is set at a 5 per cent decline per year for new plants from 1 January 2005. Note that this does not retroactively affect already-installed plants, but only new plants that have benefited from price declines in the industry.

State hand-over

As the 100,000 Roofs solar power programme succeeds, the German states are simultaneously continuing a trend of phasing out state subsidies, arguing that the federal programme provides adequate financial backing for investments. Berlin, Brandenburg, Thuringia and North Rhine-Westphalia were the only German states still offering to subsidize private PV interests in 2001. Only in Thuringia and North Rhine-Westphalia did state subsidies flow in 2002. By 2004 the German solar PV programme was completely in the hands of the federal state government.

Industrial build-up

Boosted by the REL, confidence in sustained German demand for PV systems is reflected in commercial endeavours to build up new manufacturing capacity.

All German producers of wafers, cells and modules have delivered and continue to plan substantial increases in their capacities. Production of raw silicon is planned for expansion to three industrial sites. Annual production capacity for all aspects of PV production, including wafers, cells and modules, is expanding to

$100-150MW_p$, and silicon-cell-based solar modules to over $150MW_p$. Capacity for manufacturing thin-film modules is expected to increase in the foreseeable future to $95MW_p$.

Conclusions: Size, Strength and Stability

The most recent evaluation of the 100,000 Roofs programme (KFW, 2000) provides some quite fitting conclusions to the lessons learned in Germany over the past two decades. Some of the more noteworthy conclusions are described below.

- The guarantee of feed-in payments for power produced by commercially oriented PV installations in the medium term creates an opportunity to ensure investment security in the PV industry, independent of other subsidy programmes, for the ensuing five to ten years.
- The 100,000 Roofs programme's cap of $300MW_p$ installed output for solar generators sent a negative signal to industry, which requires long-term stability to build up new industrial capacity.
- A phase-out of PV subsidies without other supportive measures, as at the end of the Thousand Roofs programme, would considerably damage existing arrangements in the market. It would lead to acute crises which threaten the existence of small and medium-sized businesses (including skilled workers and distributors) in the PV chain of production.
- A final observation is that the Renewable Energy Law, which provides a feed-in tariff high enough to operate a solar generator economically, is a very effective option to phase in grid-connected solar PV systems. A planned annual reduction of the provided tariffs forces the industry to bring down its prices, but also provides a dependable planning environment for manufacturers and developers.

Good policy-making and strong support on behalf of the German government will of course be critical to Germany's ability to maintain its position among the top three manufacturers of solar photovoltaics and in the international race for global market share in PV production. But the successes have been sufficient to create a strong industry and the failures insufficient to kill it off. And in the process, there have been many lessons from which those who follow can benefit.

Notes

1   The performance ratio of a PV system is the quotient of alternating current (AC) yield and the nominal yield of the generator's direct current (DC). It indicates which portion of the generated current can actually be used. A PV system with a high efficiency can achieve a performance ratio over 70 per cent. The performance ratio is also often called the 'quality factor'. A solar module based on crystalline cells can reach a quality factor of 0.85 to 0.95, that is a performance ratio = 85–95 per cent.

2   Cyrus Project, Greenpeace, November 1995, 'What does phasing in solar energy cost?' briefing published by Greenpeace Germany.

References

Fraunhofer Institut (1997) *SolarJobs 2010*, report commissioned for Greenpeace Germany, Fraunhofer Institut, Leipzig

KFW (2000) *100,000 Dächer-Solarstrom Programm, Statistische Kennzahlen, & Evaluierung,* final assessment report of the Thousand Roofs monitoring and evaluation program, Kreditanstalt für Wiederaufbau

11
Sustainable Energy Policy Reform in Cambodia

Andrew Williamson

Introduction

At the present time energy use in Cambodia is neither economically, socially nor environmentally sustainable. Most energy consumed is in the form of fuelwood or charcoal and almost all electricity is generated from imported fossil fuels. While Cambodia does have good resources of sustainable energy,[1] these hardly feature in official government plans for supplying the country's predicted growth in energy demand over the next 10 years and beyond. In this regard, Cambodia has much in common with many other developing countries.

This chapter aims to present a case study on sustainable energy policy reform in Cambodia, and identify lessons which may be relevant for other developing countries. It begins with relevant background on Cambodia's current political and economic situation, and the structure of its energy sector. A brief assessment of sustainable energy potential in Cambodia follows, along with a description of relevant policy, either currently in place or being considered. Finally, a discussion will reveal the people and organizations driving, or attempting to drive, the direction of energy policy in Cambodia, the barriers they face, and their successes and failures.

This chapter concludes by drawing some key lessons from the Cambodian experience of moving energy policy in the direction of sustainability. The application of these lessons may be crucial if the predicted growth in developing economies, especially in Asia, is to be achieved without massive economic, social and environmental costs.

Country Background

This section provides a brief background to relevant aspects of Cambodia's geography, recent history, politics and economy.

Geography

Cambodia covers an area of 181,035 km² in continental Southeast Asia, bordered to the east by Vietnam, to the north by Laos and Thailand, to the west by Thailand, and to the south by the Gulf of Thailand (MOE, 1994). Cambodia is approximately half the size of Japan and a third the size of Thailand. Cambodia has a population of approximately 13.8 million people, just over half of them under 18 years old.[2] Approximately 84 per cent of the population live in rural areas while most of the remainder live in Phnom Penh, the capital city, and other urban areas.

Recent history and politics

War and violence have marred most of the last 30 years of Cambodia's history. Most of the country's infrastructure and technical capacity were systematically destroyed or stolen during this time. The Khmer Rouge period from 1975 to 1979 was one of the worst periods of human destruction ever witnessed and in many respects the country is still recovering from this damage to society and economy (Kiernan, 1998). In some ways, the economy is less developed today than in the 1960s before the conflicts started (Hundley, 2003).

Cambodia is a constitutional monarchy with King Norodom Sihamoni as the Head of State since October 2004 when his father, retired King Norodom Sihanouk, abdicated the throne. The first post-war democratic elections were held in 1993 and resulted in a period of relative peace and stability. Cambodia's most recent democratic election was held in July 2003 and won by the ruling Cambodian People's Party. However, it took more than 12 months for this party to form a new government by reaching an agreement with the country's other major party.

Economy

Cambodia belongs to the group of 20 poorest nations and is classified as a Least Developed Country (Council for Social Development, 2002). More than a third of the population lives beneath the poverty line, which is equivalent to a spending capacity of US$0.50 per day; in rural areas, the proportion of the impoverished rises to 40 per cent (World Bank, 2000).

Cambodia's economy is based on agriculture, which contributes 39.6 per cent of its GDP (NIS, 2002). The garment manufacturing industry and tourism sector are playing an increasing role in the national economy. However, the country is also still highly dependent on foreign aid, which contributes 14 per cent of its GDP (NIS, 2002). This foreign aid contribution was one of the aspects threatened by the 12-month political deadlock following the July 2003 elections. Some donors halted loans and assistance packages until the government was in a position to execute binding loan agreements (Ten Kate, 2004).

Cambodia's Energy Sector

This section, which briefly describes Cambodia's energy sources, electricity supply arrangements and the parties involved, should provide some context for ensuing policy discussions.

Sources of energy

Aside from fuelwood, Cambodia has few conventional energy sources available within the country, and even fewer currently exploitable. Wood accounts for more than 80 per cent of total national energy consumption (Ministry of Industry, Mines and Energy (MIME), 1996). Fuelwood is by far the main source of energy available to the general population and is especially significant for the poor and rural dwellers. Yet, the main sources of fuelwood in Cambodia, natural forests, have been severely degraded due to widespread logging and forest land conversion for various purposes over the past 20 years (Global Witness, 2000).

For electricity production the country relies almost entirely on imported fossil fuels, mainly diesel and heavy fuel oil. There is no in-depth comprehensive geological survey data available to assess the extent of Cambodia's fossil fuel deposits. Offshore surveys of oil and natural gas have been conducted for the past 10 years with various successes and failures. Test drills have revealed the potential existence of presumably large, but as yet undetermined, offshore natural gas fields in Cambodia's portion of the Gulf of Thailand. Since neighbouring Thailand has confirmed gas deposits, and has been commercially exploiting them, the probability is high that Cambodia will in the longer term be able to undertake similar exploitation.

This would, however, require substantial investments in infrastructure and therefore confirmed domestic or foreign markets for the gas. Commercially viable offshore gas extraction will probably not be achievable for at least another five years. Three oil companies (two US and one Japanese) have signed concessionary agreements with the Royal Government of Cambodia and were scheduled to begin extensive exploration in early 2003, but the current status of activities is not known (Carmichael, 2003).

Coal deposits are thought to exist in several provinces, and offshore deposits of bituminous coal are also likely to exist. However, there has been no comprehensive national survey of either (New Energy and Industrial Development Organisation (NEDO), 2002).

Electricity supply

Electricity was introduced to Cambodia at the beginning of the last century under the French colonial administration. Today electricity is provided by a number of different organizations using many different systems, standards and levels of quality. The largest single supplier is the government-owned Electricité du Cambodge (EDC). The country's total installed generating capacity is estimated at about 411MW, with the approximate breakdown shown in Figure 11.1 and Table 11.1.

Table 11.1 *Electricity suppliers in Cambodia*

Supplier	Areas supplied	Installed capacity[3]	Annual generation
Electricité du Cambodge (EDC)	Phnom Penh plus 11 provincial towns (MIME, 2004)[4]	114MW	188.4GWh
Independent power producers (IPPs) selling to EDC	Phnom Penh and Kampong Cham (EDC, 2003)	59MW	347.3GWh
Provincial electricity operators (provincial offices of MIME)	Two provincial towns (NEDO, 2002)	3MW	5.1GWh
Rural electricity enterprises (REE) operating mini-grids	Four provincial towns and hundreds of smaller towns and villages (estimated 600 REEs) (Hundley, 2003)	60MW	No data available
Battery-charging services (REEs which do not also operate a mini-grid)	1500 battery-charging services (REEs) in hundreds of towns (Hundley, 2003)[5]	38MW	No data available
Imported power from Thailand and Vietnam (22kV lines)	11 border towns (Hundley, 2003)	21MW	No data available
Private standby diesel generation (large scale only)	All areas, but mainly Phnom Penh and Siem Reap (Hundley, 2003)	116MW	175.0GWh (NEDO, 2002)
Total		**411MW**	

Source: compiled by author from various sources

Cambodians consume on average about 55kWh per year per person, among the region's lowest rates of power consumption.

The government's electricity system consists of 24 small, isolated power systems with no transmission link between load centres. This system only reaches about 20 per cent of the population, most of them in Phnom Penh. The peak demand for EDC customers in Phnom Penh in 2003 was approximately 120MW which represents almost 70 per cent of EDC's total load and about 29 per cent of the estimated total supply capacity in the country. The peak load across all the main provincial towns is about 50MW. There were critical power supply shortages in Phnom Penh in 2004 during the dry season, February to April, when cooling loads are at their peak. This resulted in frequent rolling blackouts across the city as EDC was forced to temporarily disconnect suburbs in order to balance the system.

Electricity costs in Cambodia range from US$0.09 per kWh to US$0.53 per kWh for government services, and can be much higher for small private services or battery-charging services (Hundley, 2003). As shown in Figure 11.2, Cambodia

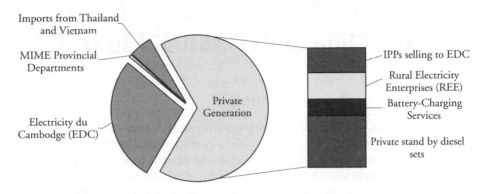

Source: Ministry of Industry, Mines and Energy, Phnom Penh

Figure 11.1 *Chart of electricity suppliers by installed generating capacity*

has the highest electricity prices of any ASEAN (Association of Southeast Asian Nations) country. This figure shows official government tariffs and excludes the higher prices of small private operators.

EDC charges a 'social tariff' of 350 Riels per kWh, which is intended to provide affordable power to its low-income, low-consumption customers.[6] This tariff has not been changed since 1995, when it was equivalent to US$0.13 per kWh. Due to depreciation of the Riel, the social tariff is now equivalent to about US$0.09 per kWh which is equal to the estimated average cost of production for EDC. However, this tariff does not cover distribution and retailing costs. In addition, EDC's costs are highly sensitive to exchange rate fluctuations since fuel and equipment purchases are incurred in US dollars. Therefore, it is clear that the social tariff is being cross-subsidized by other customers.

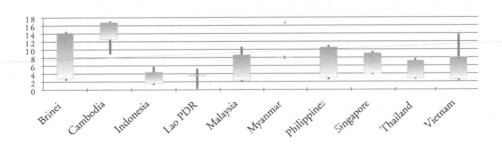

Source: ASEAN Centre for Energy www.aseanenergy.org

Figure 11.2 *Electricity tariffs in ASEAN countries for September 2003*

This cross-subsidy has some undesirable effects. First, it does not assist most of Cambodia's poor because most of them do not live in areas of the country with access to the EDC grid. Second, it acts as a disincentive to improved energy efficiency by providing power below the real cost of production and distribution. And third, it creates a barrier to the development of other sources of power that are not subsidized, such as renewable energy technologies.

It is ironic that, despite the social tariff, some of the poorest of Cambodia's urban population pay some of the highest tariffs in the country. This is because the existing laws governing EDC's power sales prevent it from retailing power to people who do not hold legal tenure over the property they occupy. A large proportion of Cambodia's poor lack legal rights to the land they live on either because they cannot afford it and are forced to squat on the land, or because property rights are still unsettled following the end of many years of conflict. In these cases, neighbouring property owners are known to purchase power from EDC at the standard tariff and then sell it on to slum-dwellers at inflated rates. While no data are available, anecdotal evidence suggests that these tariffs can be over 1000 Riels (26 US cents) per kWh, higher even than the highest official EDC tariff which foreigners must pay.

Government energy institutions

There are three Cambodian government institutions directly related to energy: the Ministry of Industry, Mines and Energy (MIME), the previously mentioned EDC and the Electricity Authority of Cambodia (EAC).

MIME has responsibility for planning and implementing the government's energy policy. The General Directorate of Energy, within MIME, consists of three departments: one for the planning of general supply and transmission options, one for the development of renewable energy and energy efficiency options, and one for the development of hydropower projects in particular. This whole directorate is almost exclusively concerned with electrical energy despite the fact that electricity is a very small fraction of the country's total energy use. MIME is a shareholder in EDC, and MIME also operates a number of mini-grids which supply some provincial towns, as explained in the previous section.

EAC was established as an autonomous government agency responsible for regulating the generation, distribution and supply of electricity, as described further in this chapter under the section entitled Policy Landscape for Sustainable Energy (see 'The Electricity Law').

Private sector: IPPs and rural electricity enterprises

Most of the power in Cambodia is supplied by private firms, as can be seen from Figure 11.1. There are at least three large commercial independent power producers (IPPs) that supply power to EDC under medium-term contracts. They all use diesel or heavy fuel-oil generators, except for one Chinese company, China Electric Power Technology Import and Export Corporation. It operates a 12MW hydropower plant built in 1968, destroyed when fighting started in 1970, and re-commissioned in 2002. Anecdotal evidence suggests the IPPs often provide additional services to EDC such as monitoring and correcting power quality because EDC lacks some of the necessary network monitoring and control equipment. The Asian Development Bank (ADB) is planning to implement a project to improve EDC's capacity in this area.

An estimated 600 rural electricity enterprises (REEs) operate small, diesel-powered mini-grids to sell power to an estimated 60,000 customers (World Bank, 2001a). The REEs are usually small, locally owned businesses that use a diesel engine and generator with low voltage distribution lines to provide power services to anywhere from 30 to 2000 local households and businesses. The average tariff charged by REEs is estimated at US$0.53 per kWh (Hundley, 2003). An estimated 1500 battery-charging businesses provide services to households and businesses, and the effective tariff is often over US$1.00 per kWh (World Bank, 2001b).

Some REEs have expressed interest in using renewable energy technologies in their businesses (SME Cambodia, 2003). A large group of REEs, mainly from provinces in the northwest of the country, attended a workshop in April 2004 to discuss the potential for biomass gasification technologies in Cambodia.[7] Consequently a project has commenced with one of the REEs to trial a biomass gasifier system to supplement diesel fuel.

The government's plans for rural electrification promote an important role for REEs, which echoes the World Bank's objectives of greater private sector participation. However, in practice, there has been friction between the public sector and private industry in some areas where the government-owned EDC has allegedly established operations in the business area of existing REEs, thus threatening the REEs' business viability. REEs are seeking longer licence periods from EAC to provide them greater security and confidence to plan further in advance and achieve investment returns over a longer period. They claim this would allow them to reduce their electricity tariffs. EAC claim they are willing to consider longer licence periods in particular cases, but are not willing to formalize such a policy. Possible reasons for this are discussed later in this chapter in the section on Barriers to Sustainable Energy Policy Reform (see 'Short leash syndrome').

The Potential for Sustainable Energy in Cambodia

An understanding of Cambodia's sustainable energy potential is a necessary basis to discuss its role in Cambodia's energy mix, and the effectiveness of current policy. However, there is no official and comprehensive review of Cambodia's renewable energy resources or energy efficiency opportunities. This presents a significant barrier to sustainable energy development, and may lead to energy policy that ignores its potential contributions and benefits to the country.

The author has conducted an initial desk-based assessment of Cambodia's potential for sustainable energy development (see also Williamson et al, 2004). For simplicity this exercise excluded energy related to transport. The findings are summarized in Table 11.2, forming a foundation and providing some perspective for the discussions in the following sections. However, care should be taken with

Table 11.2 *Summary of the estimated potential and status of sustainable energy generation and savings in Cambodia*

	Technical potential (GWh/year)	Currently installed projects (GWh/year)	Theoretical remaining potential (GWh/year)	Potential annual greenhouse gas abatement (kilotonne CO_2 equ)
Hydropower	37,668	55	37,613	26,228
Biomass	18,852	0	18,852	13,146
Solar	65	1	64	44
Wind	3665	0	3665	2556
Industrial energy efficiency	547	0	547	381
Residential energy efficiency	6591	29	6562	4576
Total	67,388	85	67,303	46,931

Source: Williamson et al (2004)

the use of these results because they are necessarily approximate due to limited data and the limited scope of this assessment.

These results were based on estimates of Cambodia's renewable energy resources and energy efficiency opportunities provided by a number of different studies.[8] The calculations do not consider the financial feasibility for developing this renewable potential, as this will change over time. However, the results do provide a useful baseline to understand the available potential with respect to the country's energy needs.

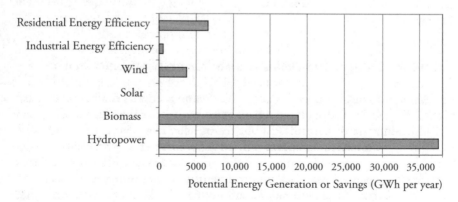

Source: Williamson et al, 2004

Figure 11.3 *Estimated sustainable energy potential in Cambodia*

The simple analysis above provides some profound insights into the magnitude of sustainable energy resources available to Cambodia. This exercise estimates that 7000GWh of energy could potentially be saved through improved generation and end-use efficiency, mainly in the use of wood. This is equivalent to more than 13 times the amount of electricity EDC supplies to all its customers.[9] This analysis also estimates that about 60,000GWh per year of energy could be generated from Cambodia's renewable energy resources; almost three times the total energy consumed by the whole economy.[10] Renewable energy projects already identified and studied in some detail, mostly hydropower projects, offer a potential output of about 9000GWh per year.

Clearly significant renewable energy sources are available for exploitation in Cambodia, and there is also a significant role for increased energy efficiency. The remainder of this chapter will discuss the policies and approaches being used to promote the development of this potential.

The Policy Landscape for Sustainable Energy

The Cambodian government does not have any policy to specifically promote the use of renewable energy or energy efficiency. This section will look at the current energy policy landscape with respect to the support, or lack thereof, offered to sustainable energy development. It covers official government policy as well as studies and recommendations by a variety of actors interested in influencing government policy.

National energy sector development policy

The Cambodian government established a National Energy Sector Development Policy in 1994. This policy sets a very broad, generic framework for the development of laws and practical policy regarding the country's energy sector. Consequently it is technology neutral and does not specify any particular design methodology.

This policy was drafted at a time when Cambodia had only just recently achieved peace and democratic rule. So one might assume this policy would focus on development at all costs through the supply of the cheapest power possible. However, the policy adopts a surprisingly holistic and long-term view.

This is demonstrated by the following summary of the policy's intent:

- to provide an adequate supply of low-cost energy for homes throughout Cambodia;
- to ensure a reliable, secure electricity supply at prices which encourage investment in Cambodia and economic development;
- to encourage the exploration and development of environmentally and socially acceptable energy resources that are needed to supply all sectors of the Cambodian economy; and
- to encourage efficient use of energy and to minimize environmental effects resulting from energy supply and use.

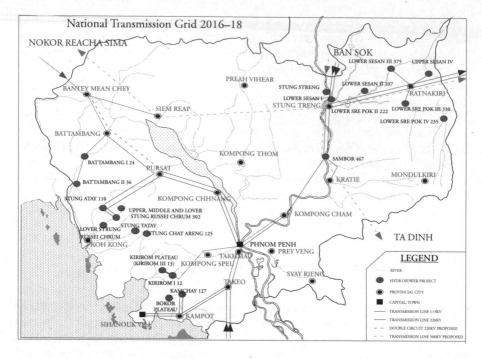

Source: MIME (2004)

Figure 11.4 *Planned generation and transmission system*

So the foundation of Cambodia's energy policy seems quite appropriate to support the development of renewable energy and energy efficiency opportunities. This provides a very useful starting point for the development of policies and initiatives for sustainable energy. However, this policy is certainly not sufficient in itself, as will soon become apparent from discussions in this chapter.

Energy master plans

World Bank

In 1998 the World Bank commissioned a consultancy firm to develop a power transmission master plan and rural electrification strategy for Cambodia. The master plan covers load forecasts, transmission development and power generation options. In brief, the plan advocates the development of high voltage transmission lines across the country to allow the import of power from neighbouring countries, plus the development of a series of large hydropower schemes involving dams on a number of rivers including the Mekong and some if its tributaries. The report briefly dismisses any potential contribution from renewable energy sources, apart from hydropower, as discussed in this chapter under the section entitled 'Energy Policy Reform Actors in Cambodia' (World Bank and HECEC, 1998).

The government accepted the consultants' recommendations and adopted a summary of their report as the official Cambodia Power Sector Strategy 1999 to 2016 (MIME, 1999). MIME is currently producing a Cambodia Energy Sector Strategy which is based heavily on the 1999 document, but with updated plans and a new section on wood energy policy (MIME, 2004).

New Energy and Industrial Technology Development Organization (NEDO)

In 2001, NEDO of Japan embarked on a study to support the creation of an energy master plan for Cambodia. Their report includes an assessment of renewable energy potential in Cambodia (NEDO, 2002). NEDO's study provides an interesting alternative to the government's official plans. Like the government plans, NEDO recommends building a national grid network; but only extending to areas of major population and industry. Outside of these areas, NEDO supports renewable energy-based mini-grids for rural electrification. This proposal is based on a relatively thorough analysis of the potential for renewable energy in Cambodia.

Unfortunately, the study for this NEDO energy master plan was completed some three years after the adoption of the master plan funded by the World Bank. The proposed NEDO energy master plan appears to have had no impact on the national transmission and rural electrification strategy.

The Japanese International Cooperation Agency (JICA) offered to send a team of Japanese engineers to Cambodia for a period of two years, commencing in late 2004, to produce a Renewable Energy Master Plan. The scope of the plan, and how it will differ from the NEDO study, is unclear at this stage.

The Electricity Law

The Electricity Law of the Kingdom of Cambodia was established in 2001 to govern the operations of the electric power industry and the activities of licensees that provide electric power services, including:

> *The principles for: (i) the protection of the rights of consumers to receive the reliable and adequate supply of electric power services at reasonable cost, (ii) the promotion of private ownership in the facilities for providing electric power services, and (iii) the establishment of competition wherever feasible within the electric power sector.* (Article 2)

The Electricity Law also established the EAC as an autonomous public body to regulate electric power services. The duties of EAC include the issuance, revision, suspension, revocation or denial of licences for the supply of electricity services. The law also gives EAC the duty of ensuring that the provision of services is performed in a 'transparent manner' and that the public be informed about affairs within its duties.

EAC regulates tariff rates and in 2004 the ADB funded a project to establish new regulations, guidelines and procedures for tariff setting in Cambodia. At the time of writing this study, these documents were in draft form and had been presented for comment at an industry seminar.

The proposed new process for tariff regulation appears to be a good move for achieving consistency and equity across the industry, plus encouraging investment and long-term tariff stability. However, the process may also introduce some new barriers to the use of renewable energy technologies. This is because it proposes to establish a tariff table listing the maximum allowable tariff for each type of licensee. If a licensee disagrees with the maximum tariff, it may submit a request to EAC for a special tariff review.

The review process which EAC must follow involves analysing the REE's business costs and determining if they are 'reasonable' in terms of supplying efficient and affordable power services. Unfortunately, the proposed review process does not require EAC to consider any issues other than the tariff. This could prevent the development of renewable energy projects which often require a higher tariff initially, but which can be cost competitive over the life of the equipment. The additional benefits of these projects, such as improved energy security, tariff stability and reduced pollution, would not be considered by EAC's tariff review.

Neither the Electricity Law nor the proposed tariff regulations make any mention of renewable energy or energy efficiency. In theory, this should not hinder the development of sustainable energy so long as the policy framework is suitable. However, in practice, it seems that having an energy policy with good intentions is not sufficient to see them incorporated effectively in regulation.

Renewable Electricity Action Plan (REAP)

The World Bank hired an expatriate consultant to produce a REAP for Cambodia to encourage the generation of electricity from renewable energy sources. The first edition of the document was produced in May 2003 with assistance from government MIME staff and the input of 150 industry stakeholders, both local and foreign, through a series of national workshops over a period of almost three years (Hundley, 2003).

The REAP sets the following industry development targets and proposes activities and budgets to achieve them:

*   6MW of installed capacity from renewable energy;
*   100,000 households provided with electricity services;
*   10,000 solar home systems installed; and
*   the creation of profitable, demand-driven markets for renewable energy-based electricity.

The REAP proposes a partnership between public and private sectors to facilitate investment in renewable energy projects, in particular hydropower. The private sector would be left to develop and supply the projects, while the government would ensure suitable market conditions.

The REAP is endorsed by MIME, but has not been adopted as official government policy. However, a summarized version, which proposes a basic five-year action plan, has been prepared for the government's consideration. While the REAP has support within MIME, there are no obvious plans for its implementation, nor is there a donor forthcoming with the estimated US$50 million required to fund it.

National policy and strategy for renewable energy

The World Bank commissioned an international consulting firm in 2003 to prepare strategy and policy documents for renewable energy-based rural electrification in Cambodia. This activity was a follow-up to the REAP, and was seen as an important precursor to the Rural Electrification and Transmission project. This project will be implemented with loan funding from both the World Bank and the ADB, and a grant from the Global Environment Facility. This is an extensive project with activities ranging from building a transmission line from Vietnam to Phnom Penh for importing power, to the establishment of a Rural Electrification Fund (REF) to support the development of rural mini-grids, especially those based on renewable energy generation.

The proposed policy document is based on the government's 1994 energy policy. The sustainability focus of the 1994 policy is expanded in this document into six specific intentions of the policy, and then a number of objectives and guidelines for achieving them.

The strategy document briefly describes a number of activities needed to achieve its targets. These activities include implementation of the REAP and establishment of the REF, and encouragement of private sector involvement. The strategy document ends by setting the same targets as stated in the REAP, which are listed above. These strategy and policy documents have not been adopted by the government, and there are no obvious plans to do so.

Conclusions on the sustainable energy policy landscape

Cambodia's National Energy Sector Development Policy from 1994 established a good basis for sustainable energy developments. Significant time and resources have been devoted since then to ensure that the government's plans and strategies are consistent with the original policy's support for sustainable energy. However, this appears to have had little impact to date, as sustainable energy development does not feature in any of the government's official plans for the sector.

Energy Policy Reform Actors in Cambodia

Policy reform initiatives in any country are often the result of a number of different lobbying interests within the country. In the case of a developing country like Cambodia, policy reform is greatly influenced by external interests such as foreign donor countries and multilateral donor agencies. This section looks at who is driving the major energy policy reform initiatives in Cambodia, the techniques they use, and their effectiveness.

Donor agencies

Foreign donor agencies are powerful policy reform actors in Cambodia. They usually achieve their policy aims by providing the relevant government ministry with the resources and technical expertise needed to develop and draft policy documents. This usually consists of a team of foreign consultants and a series of national conferences. This approach gives a donor significant influence as Cambodia's ministries are usually completely lacking in such resources and expertise.

Some donor-led policy initiatives have already been described here, such as the ADB's new tariff-setting regulations and NEDO's study for an Energy Master Plan. Another example is a project funded by JICA, which hired a team of Japanese engineers for more than a year to draft a set of technical standards for electrical power. These standards will, theoretically, be enforced and used to govern electricity generation and distribution in Cambodia. However, in practice it will be a very difficult task to enforce such standards in Cambodia and it is currently beyond the means of the ministries responsible for such tasks.

This last example serves to highlight an issue regarding the effectiveness of donor-led policy reform. While some donors have proved to be highly effective in having new policies adopted by government, the real effectiveness of policy reform must be measured by its implementation and enforcement. This appears to be harder for donors to influence, as will be discussed towards the end of this chapter under the section entitled 'Barriers to Sustainable Energy Policy Reform'.

The World Bank

Following the years of conflict and war in Cambodia, the World Bank has probably been the single most active player in the nation's energy sector. It has been very effective in driving government policy and thus its role in energy policy deserves special attention here.

The World Bank's involvement in Cambodia's Power Sector Strategy 1999–2016 and the REAP was described earlier in the previous section. These are both good examples of the World Bank's policy reform activities, which were achieved by arranging and funding the development and implementation of the desired policies. Its consultants undertook the activities which are usually left to governments, including background studies, stakeholder engagement and drafting of policy documents.

These two examples display very different policy agendas. The energy master plan, and the study behind it, displays a conservative energy planning mentality based on large central generation assets and national transmission infrastructure. In fact, as mentioned earlier, the consultant's report rejected any potential for renewable energy or energy efficiency and only focused on large hydroelectric dams (World Bank and HECEC, 1998). However, the second initiative for the REAP was quite the opposite, with the outcome advocating very strongly a decentralized approach to energy supply in rural areas based on renewable energy resources, and mainly implemented by an army of small private entrepreneurs (Hundley, 2003).

The different approach of these two exercises is more reflective of changing World Bank philosophy than any indication of a change in Cambodian government priorities.

Private interests

Some private firms and NGOs have played a role in sustainable energy policy development in Cambodia. Probably the most notable example is the influence that the association of REEs has had with respect to licensing and tariff issues. In the new draft tariff-setting regulations, EAC has gone to great lengths to establish special classifications and procedures suited for REEs as distinct from large licensees such as EDC.

This is an indication of a big improvement in the profile of REEs and awareness of their issues among policy-makers. The formation of an association of REEs over a year ago precipitated this change. The association raised the profile of their issues in 2003 when it published a confrontational letter to EAC bluntly describing the inequities and risks faced by REEs, as described earlier in this chapter (see 'Private sector: IPPs and rural electricity enterprises').

A period of tension between the government and REEs ensued but eventually appeared to subside. The particular mention of REEs as a valid and important class of licensee in the new tariff-setting regulations is a sign that the association's moves may have had some impact.

While the REE association has not acted on any policies related to sustainable energy, this may soon change if REEs start to develop renewable energy projects. The REE trial of a small biomass gasifier described earlier suggests this is likely. There has also been some interest among REEs about the possible introduction of a net-metering policy in Cambodia which would establish standard procedures and prices for small grid-connected generators to sell power back to EDC. This would be one solution that could allow REEs to maintain their business in the event of the public grid being extended to their franchise area. Such net-metering policies have proved an effective way to encourage renewable energy in other countries because they can improve the financial feasibility of grid-connected projects.

One notable example of a particularly unsuccessful policy reform effort in Cambodia is the quest for the removal of import duties on renewable energy equipment. The current tariffs are a 35 per cent import duty, plus a 10 per cent value added tax which is calculated on both the cost of the equipment plus the duty, so the total tax paid is effectively about 48 per cent. The mainly small, private businesses that import this equipment into Cambodia have been advocating the importance of this policy change for at least three years. Recently the World Bank has supported their cause by urging the government to reconsider the duties before the start of the REF so as to prevent, in effect, the REF subsidies simply funding payment of the import duties to the government. However, there is still no indication that the government is seriously considering changing the duties.

The government

The Cambodian government often appears to be a willing partner in policy reform for greater sustainable energy development. The eagerness of most ministries to cooperate with donors to develop and implement new energy policies has already been described here.

This support appears to start at the top with the Prime Minister of Cambodia, Mr Hun Sen, who has recently emerged as an unlikely advocate for greater renewable energy development. During the political campaign leading up to the national election in July 2003 he made regular media appearances at opening ceremonies for new solar power systems installed at various schools, pagodas, bridges (with solar-powered lighting) and health centres across the country. These systems were apparently all gifts made by the Prime Minister using his own personal funds. He also recently issued a challenge to his newly-elected government to put their money where their collective mouths are and support rural development by donating solar power systems to communities just as he had. He proposed that, if every minister and senior bureaucrat donated a system to one school, pagoda or health centre every year, a significant impact could be made in a short time.

Such high-level support is a good start for Cambodia, and would be a welcome example to be followed by leaders of many developed countries. However, in Cambodia as elsewhere, political rhetoric is not always followed by consistent regulatory action.

Conclusions about the policy reform actors in Cambodia

Cambodia's experience highlights the power of influence large donors such as the World Bank have over energy policy in developing countries. This offers the advantage that, assuming the donor's objectives are aligned with the country's best interests, policy reform can be achieved relatively quickly. However, the disadvantages include the risk that policy is imposed on a country without an appropriate level of public debate. This may mean that the government does not share the donor's level of commitment to the policy, understanding of its implications, or resources to implement and enforce it effectively.

The failure of both the World Bank and small private interests to change the existing import duties on renewable energy equipment suggests another important lesson. It suggests the techniques and resources of the policy reform actor may be irrelevant in cases where the desired policy reform threatens a government's source of income or political power.

Barriers to Sustainable Energy Policy Reform

Barriers to the greater use of sustainable energy in Cambodia are widely acknowledged and often cited. These barriers are similar to those found in many developing countries, and some are also shared with developed Western countries. However, what is not so widely understood are the often subtle barriers to the development and implementation of sustainable energy policy in Cambodia. This section discusses some of these barriers.

Technology stigma

Renewable energy technologies are widely considered to be a nice idea, but an expensive luxury more appropriate for rich Western countries than for developing ones. In fact proposals for wider renewable energy use in developing countries often attract the same stigma surrounding any issue that weighs up concerns about environmental sustainability against economic development. The common basis of this stigma, stated bluntly, is that it is unfair for rich environmentalists from the developed world to urge developing countries not to harm their environment, as they strive to grow, through exploiting their natural resources.

This stigma may have been established, or at least reinforced, by technical advisers from donor countries and agencies. This is because in most Western countries renewable energy is still very much a luxury. There are usually much cheaper sources of energy available, and therefore the only reason to invest in renewable energy technologies is if specific environmental concerns are a higher priority than financial ones.

However, this logic does not always apply in developing countries where conventional fossil fuel-based electricity is extremely expensive. For example, the average rural tariff in Cambodia of about US$0.50 per kWh would be sufficient to finance almost any renewable energy technology. Compare this tariff, for example, to private commercial wind farms being developed in Australia based on a power price of around US$0.05 per kWh. Though it is not valid to compare these prices directly without considering other important factors such as cost and availability of project finance in Cambodia, existing infrastructure, and the sovereign and contractual risks involved, the vast price difference serves to illustrate the point that Western attitudes towards the relative cost of renewable energy technologies are not always applicable in developing countries. Yet this resulting stigma against the use of clean energy technologies in Cambodia creates a subtle but effective barrier to sustainable energy policy development.

Lack of data

The lack of data on Cambodia's renewable energy resources and energy efficiency opportunities was discussed earlier in this chapter (see 'The Potential for Sustainable Energy in Cambodia'). This data void creates a barrier to project development, but one that can be overcome with sufficient time and money.

However, effective policy development requires more data than simply knowledge of the extent of renewable energy resources. Policy-makers need access to current accurate statistics on electricity consumers and suppliers, and business costs for project development, operation, maintenance and overall profitability. None of this data is readily available in Cambodia. One reason is that the relevant institutions lack sufficient resources and technical capacity to collect the data. Another is that there is a strong distrust of government among private businesses, especially relating to potentially sensitive business data which could be used to extract higher taxes and fees, or assist competitors.

Short leash syndrome

Policy-makers must take into account the enforcement methods available in a particular sector if their policies are to be effective. It may be futile to develop policies and regulations that require intensive enforcement if this is not realistically available. Most government agencies in Cambodia tasked with enforcing the government's regulations struggle due to a severe lack of funding and technical capacity. In this situation it is easy to understand how enforcement officials may be vulnerable to corruption and left unable to perform their job effectively.

Issuing short-term licences is one pragmatic approach to this problem because it effectively forces licensees to approach the regulator for review and approval at regular intervals. This clearly provides the regulator with more opportunities to raise revenues from licence fees. However, it may also allow cheaper and more effective enforcement because it is easier to refuse to renew a licence, or to change licence conditions before re-issuing, than to investigate and reprimand an errant licensee.

This preference for regulators to keep licensees on a 'short leash' can present a barrier to implementing effective policies for sustainable energy development which generally require long-term conditions to foster investor confidence.

High hopes

A recurring feature of each energy master plan described earlier (see 'The Policy Landscape for Sustainable Energy') is that Cambodia's current energy problems will be solved in the medium term once high voltage transmission lines are built to Thailand and Vietnam. The theory goes that initially these links will allow cheap power to be imported, and that further down the line Cambodia will export its own cheap power generated from large hydropower projects. An additional source of income is expected to materialize once Cambodia's small offshore oil and gas reserves are developed.

The assumptions behind these claims are questionable. Power purchase agreements have already been negotiated for the import of power from Thailand and Vietnam, and so it may be some time before Cambodia's status is changed from buyer to seller. Furthermore, analysts are beginning to warn of the potentially disastrous economic and social effects that can result from the sudden flow of oil and gas revenues into a small developing economy without effective governance. The situation in Nigeria is one example of this: inflation has spiralled, corruption is endemic and the general population appears not to have benefited from the exploitation of its fossil fuel reserves (Carmichael, 2003).

Despite the debatable accuracy of these high expectations, they may nonetheless be having a subtle effect on Cambodia's energy policy development. Put simply, there is little motivation for policy-makers to develop sustainable energy policies if an expectation exists that an energy solution will soon appear in the form of cheap power from their neighbours, along with both cash and cheap fossil fuels from their own backyard.

Conclusion

Cambodia is one of the world's poorest countries, yet its citizens pay some of the world's highest prices for commercial energy. Transport and power production rely almost entirely on imported fossil fuels, and households still rely on energy from wood. However, the country has some good renewable energy resources, the development of which could help reduce poverty and improve energy security and general sustainability.

In many ways the conditions for sustainable energy development in Cambodia are ripe for development. The cost of power from other sources is extremely high and there is hardly any existing infrastructure, so decentralized generation is often a good option. However, a number of significant barriers must be addressed through the development of effective policy.

This chapter has raised a number of issues regarding the development of effective policies for sustainable energy development in Cambodia. Some lessons emerge that may be relevant for other developing countries and Western organizations working there.

In the case of Cambodia, policy for sustainable energy development can be acquired quite easily although at a cost; that is to say, it is possible to hire consultants to draft policy and have it adopted by government. The success of this approach in other developing countries cannot be assumed. Countries with greater technical capacity and tighter governance may resist attempts by Western donors to influence their policies.

However, it is also clear that such an approach does not always guarantee that the implemented policies will have the desired effects. This requires high-level government commitment to support, promote and enforce the new policies. Achieving this commitment is often more difficult and less predictable than simply having a policy document adopted.

Formulating and implementing sustainable energy policy in a developing country like Cambodia provides a unique set of challenges and opportunities. Many challenges are due to foreign influences, rather than inherent failings of the government. This means they can be more easily managed; good news to anyone interested in developing sustainable energy options in the developing world.

If one common lesson can be extracted from this discussion, it is that the low resources and technical capacity of a country like Cambodia can offer significant opportunities for positive change. The extent and speed of this change can exceed comparable change in a developed country. In this regard foreign donors can wield great power for achieving policy reform, but this power requires responsibility to maximize the positive potential benefits for the country concerned.

Developing countries like Cambodia, especially those in Asia, may represent a significant proportion of the world's predicted energy demand and economic growth in the future. This provides a big opportunity to improve the lives of millions who are now in poverty. However, if effective policy is not established in the near future to ensure some level of sustainability, then the opportunities to ensure sustainable growth may be lost.

Acknowledgements

The author is very grateful for the generous support and assistance of Bridget McIntosh, Tanak de Lopez, Tin Ponlok, Matthew Coghlan, and the staff of the Ministry of Industry, Mines and Energy in Phnom Penh.

The Cambodian Research Centre for Development (CRCD) is an independent, non-political, non-partisan and not-for-profit research organization. The activities of CRCD focus on research in the areas of development, with the goal of improving the activities of national and international organizations involved in the development of the Kingdom of Cambodia. CRCD aims to provide a broad public knowledge for sustainable development, through academically rigorous, non-partisan and independent research. More information is available at www. camdev.org.

Notes

1 For the purposes of this chapter the term 'sustainable energy' is used in a general sense to refer to both renewable energy and energy efficiency technologies, practices and policies.
2 These figures are a projection by the National Institute of Statistics of Cambodia's population in 2003, based on the 1998 national census.
3 The installed capacity figures are estimates from different sources and for periods from 2001 to 2004.
4 The power generation and distribution systems in eight provincial towns were upgraded under a project funded by the Asian Development Bank (ADB) and transferred to EDC in March 2004.
5 Estimated total installed capacity of the estimated 1500 battery-charging services, assuming an average generator size of 25kW.
6 At the time of writing, September 2004, the exchange rate was US$1.00 = 3900 Cambodian Riels.
7 This workshop was organized by Small and Medium Enterprise Cambodia (SME), an NGO based in Phnom Penh.
8 The main reports used were: TrueWind Solutions, 2001; NEDO, 2002; Meritec, 2003; and CFSP, 2004.
9 EDC's Annual Report for 2002 claims its total generation was 535.703GWh for the year.
10 Cambodia's Energy Balance for 1995, issued by MIME, estimates the country's total consumption of all types of commercial energy to be 77,720.87TJ which is equivalent to 21,589GWh.

References

Carmichael, R. (2003) 'A volatile, high octane blend; Oil and gas don't mix well with corruption and weak governance', *Phnom Penh Post*, 15–28 August, pp8–10
CFSP (2004) 'Legal and sustainable charcoal', paper presented by Cambodia Fuelwood Saving Project to the Improved Cookstove Workshop, Phnom Penh, 25 February
Council for Social Development (2002) *National Poverty Reduction Strategy 2003–2005*, Council for Social Development, Phnom Penh
Electricité du Cambodge (EDC) (2003) *Annual Report 2002*, Electricité du Cambodge, Phnom Penh
Global Witness (2000) *Chainsaws Speak Louder than Words*, Global Witness, London
Hundley, C. (2003) *Renewable Electricity Action Plan*, World Bank, Washington, DC
Kiernan, B. (1998) *The Pol Pot Regime: Race, Power, and Genocide in Cambodia Under the Khmer Rouge, 1975–79*, Yale University Press, New Haven, CT
Meritec (2003) 'Pre-investment study of community-scale hydro projects, Cambodia', report by Meritec Ltd for the New Zealand Ministry of Foreign Affairs and Trade, Auckland
Ministry of Environment (MOE) (1994) *First State of the Environment Report 1994*, Ministry of Environment, Phnom Penh

Ministry of Industry, Mines and Energy (MIME) (1996) *Strengthening Energy Policy in the Department of Energy Planning*, MIME, Phnom Penh

MIME (1999) *Cambodia Power Sector Strategy 1999 to 2016*, MIME, Phnom Penh

MIME (2004) *Cambodia Energy Sector Strategy – Draft for Comment*, MIME, Phnom Penh

National Institute of Statistics (NIS) (2002) *Cambodia Statistical Yearbook 2002*, Ministry of Planning, Phnom Penh

New Energy and Industrial Development Organisation (NEDO) (2002) *Assistance Project for the Establishment of an Energy Master Plan for the Kingdom of Cambodia – Final Report*, NEDO, Tokyo

SME Cambodia (2003) *Rural and Renewable Electricity Project*, SME Cambodia, Phnom Penh

Ten Kate, D. (2004) 'ADB says no to new loans without assembly', *Cambodia Daily*, 11 February, p17

TrueWind Solutions (2001) *Wind Energy Resource Atlas of South East Asia*, World Bank, Washington, DC

Williamson, A. Delopez, T. McIntosh, B. and Tin, P. (2004) *Sustainable Energy in Cambodia: Status and Assessment of the Potential for Clean Development Mechanism Projects*, CRCD and Institute for Global Environmental Strategies (Japan), Phnom Penh

World Bank (2000) *Country Assistance Strategy for the Kingdom of Cambodia*, World Bank, Washington, DC

World Bank (2001a) *Final Report on RE Strategy and Program for the Rural Electrification Strategy and Implementation Program for the World Bank*, Meritec Ltd, Phnom Penh

World Bank (2001b) *Market Study and Project Pipeline Development for Solar Photovoltaics*, Burgeap and Kosan Engineering, Phnom Penh

World Bank and HECEC (1998) *Kingdom of Cambodia – Power Transmission Master Plan and Rural Electrification Strategy*, World Bank, Phnom Penh

Index